dixi
books

Visions for a Post-Covid World:

Defining a Radically New Normal

Edited by Joe Gray

Dixi Books

Visions for a Post-Covid World: Defining a Radically New Normal
Edited by Joe Gray
Proofreading: Andrea Bailey
Designer: Pablo Ulyanov
I. Edition: March 2021

Library of Congress Cataloging-in Publication Data
1st ed.
ISBN-13: 978-1-913680-11-4
1. Rewilding 2. Sustainability 3. Economics 4. Education
5. Environmental Conservation 6. Energy 7. Food Systems
8. Climate 9. Animal Rights 10. Communication

© Dixi Books Publishing
293 Green Lanes, Palmers Green, London, England, N13 4XS
info@dixibooks.com
www.dixibooks.com

Visions for a Post-Covid World:

Defining a Radically New Normal

Edited by Joe Gray

The Voice of the New Age

Contents

Introduction

Joe Gray

"There comes a time when humanity is called to shift to a new level of consciousness... That time is now."
Wangari Maathai

We are living through a period of extraordinary global campaigns and the corresponding promise, if one has any optimism, of extraordinary change. Recent years have seen huge numbers of people across the globe, and from various sectors of society, showing their support for the Extinction Rebellion and Black Lives Matters movements – sometimes in ways that have put individuals' freedom and safety at risk. Meanwhile, school and university students have staged a multitude of Fridays for Future climate walk-outs.

These movements, and others like them, are united by a desire to create a world that is fairer – for humans and for all life. And make no mistake. Slavery may have been outlawed, women's suffrage introduced, and protection for endangered species legislated; but injustice still remains the hallmark of modern humanity.

And then, with ever-more people clamouring for change, something happened that forced change upon us at a scale and speed that caught the world by surprise. That thing was a global pandemic.

The impacts of Covid-19 have already been enormous. The virus has caused physical suffering and mental distress, brought many lives to an early end, and put an incredible strain on modern health and welfare systems and broader economies.

At the same time, it has made industries find ways of working that, where feasible, no longer necessitate a daily mass transportation of people. The outcomes of this specific transition may be far from perfect, but it is hard not to be amazed by the way in which it has unfolded.

The reduction in travel for work and other reasons has given many individuals living in societies where such movement is the norm an opportunity to be exposed to the benefits of being in place. These include exploring local green spaces that might previously have been overlooked, connecting with neighbours through mutual support networks, and enjoying the non-homogenised and personal shopping experience provided by local stores. Conversely, for people who have suffered from social isolation or been trapped in households where domestic violence prevails, the restrictions on movement may have been a brutal ordeal.

Perhaps most significantly of all, in the long run, the virus has shone a light on some of the greatest failings of modern humanity and given us cause to deeply appraise our current course. To be clear: with a still-growing population and ever-worsening life-support systems, humanity is headed for suffering on a scale that will eclipse that witnessed during the pandemic by some orders of magnitude. Moreover, we are currently bringing the rest of the living world down with us, like a drowning person pushing another individual under the water in a desperate attempt to find oxygen – but on an almost infinitely larger scale.

Change of an extraordinary nature is urgently needed. And that change must find value in all life. Only when humanity advances to a stage where other beings are seen no longer as commodities or resources, but as individuals with meaningful lives in their own right, will we have given ourselves and all our Earthly neighbours a genuine hope of a just future.

Our focus in the present book is harnessing the current hunger for change and looking at how, beyond Covid-19, the world can become a better place for humans and non-humans alike.

In some ways, I am not the right person to be writing this Introduction. My own response to the media tsunami arising from the pandemic has been to mostly shut myself off from the news and retreat into a bubble where the reigning concerns have been local in nature. I have even written a book about my experiences of life in the tiny garden of my suburban home during the first 'lockdown' (Gray, 2021). Has my response been a coping mechanism? Almost certainly. Is it something of which I am ashamed? Absolutely not.

On occasions when I have temporarily emerged from my self-imposed blackout, resensitised by the withdrawal, I have found emerging news stories about the state of the world around us to be starker than ever. For instance, the latest *Living Planet Report*, published in September 2020, noted the following declines between 1970 and 2016 in the combined population sizes of mammals, birds, amphibians, reptiles, and fish: 24% in Europe and Central Asia, 33% in North America, 45% in Asia Pacific, 65% in Africa, and 94% in Latin America and Caribbean (WWF International, 2020).

I'm going to restate that last item twice, because I had to read it three times to convince myself that my eyes were not deceiving me. *In less than half a century, we have reduced vertebrate populations in the tropical subregion of the Americas by nearly 95%. In less than half a century, we have reduced vertebrate populations in the tropical subregion of the Americas by nearly 95%.*

Then, a few months later, a paper in *Nature* reported that 2020 was the year in which global human-made mass overtook living biomass in extent (Elhacham et al., 2020). As Eddie Vedder muses in the Pearl Jam song *Garden*: "I don't question our existence, I just question our modern needs." And if ever there was an event that highlighted the utterly irresponsible scale of modern society's perceived 'needs', this crossing-over point was surely it.

oOo

Introduction

With the present book, I'm delighted to have seen writing from so many great thinkers in sociological, environmental, and ecological fields coming together – after all, the solutions to society's deep problems will necessarily be multi-faceted. Equally, I am humbled to have had the opportunity to contribute a chapter myself – with the inspirational Reed Noss as co-author, no less – and to pen this Introduction to the collection. I cannot take credit for the finer details of the editing process, as the publisher has put in the lion's share of the work in that regard. Instead, I have been able to focus on the more enjoyable tasks of inputting into the selection of topics and authors and broadly shaping what we hope is a coherent whole.

Below, I provide a brief overview of each of the chapters in this book. First, I should note here that, as well as creating the impetus for this collection, Covid-19 has become intertwined with the editing process itself. A couple of authors who we dearly wanted to make contributions were, in the end, unable to write a chapter on account of the challenges of balancing this with the increased pressures that have arisen from life during a pandemic. We wish them, and all others who are struggling during these difficult times, the very best.

Chapter-by-chapter overview

Chapter 2: Anja Heister looks towards a post-Covid-19 future in which we learn to view other animals as equals and where we pursue liberty, freedom, and happiness together.

Chapter 3: George Wuerthner explores why striving towards protecting half of the Earth or more under strong conservation measures is not only an achievable goal but an essential one for an ecological civilisation.

Chapter 4: In the first of two case studies, members of the Karen Environmental and Social Action Network describe the Salween Peace Park in Myanmar, an Indigenous-led initiative whose goal is to prevent destructive development and promote peaceful coexistence with non-human beings. Shining examples like these continue to remind the world of why we need to listen to the wisdom of Indigenous voices.

Chapter 5: Reed Noss and I argue why a major revival of natural history, as the "practice of intentional, focused attentiveness and receptivity to the more-than-human world," is urgently needed, and we offer suggestions for how such a revival can be brought about.

Chapter 6: In our second case study, we present extracts from an article published by *The Conversation* on an early-childhood programme being taught in forests and meadows on the banks of the GabeKanang Ziibi (the Humber River), outside the Canadian city of Toronto. The programme promotes Indigenous wisdom, starting with the Anishinaabe teaching that *Nibi* (water) is the blood of *Aki* (the Earth).

Chapter 7: Peter Gray, a critic of modern compulsory schooling, describes an educational evolution that began before Covid-19 and may, he argues, be hastened by the pandemic into a revolution.

Chapter 8: Andrew Olivier appraises the state of the working world and champions eco-communities as an "alternative future" of sustainable living that is already here.

Chapter 9: Colin Tudge presents an impassioned rebuttal of the misinformation about food and farming that is spread by "the government, the corporates, big finance and large, but mercifully not all, sections of academe," and he shows why, for better agriculture, we must take matters into our own hands.

Chapter 10: Marie Claire Brisbois examines how energy-use patterns have changed during the pandemic and envisions a decentralised 'energy democracy' in which citizens have taken back power, in both senses of the word, from the corporate behemoths.

Chapter 11: Martin Hultman rallies against climate change denial and the ecocidal logic of the Anthropocene, and he calls for a focus on ending the destructive idea of human exceptionalism.

Chapter 12: Sibylle Frey, Max Winpenny, and Brittany Ganguly present their ideas for improving the way in which environmental ideas are communicated, so that positive action can be inspired and societal transformation achieved.

Chapter 13: Eileen Crist calls for humanity to secede from a "dominant no-limitations, life-destroying" civilisation and build alternative communities on "rational and sacred ground."

Closing remark

That is more than enough from me for now, and so I will give the last words to Wangari Maathai (1940–2011), an activist from Kenya who was instrumental in the founding of the Green Belt Movement and whose thoughts also opened this Introduction:

"Today we are faced with a challenge that calls for a shift in our thinking, so that humanity stops threatening its life-support system. We are called to assist the Earth to heal her wounds and in the process heal our own – indeed to embrace the whole of creation in all its diversity, beauty and wonder."

References

Elhacham, E., Ben-Uri, L., Grozovski, J., Bar-On, Y. M., and Milo, R. (2020). Global human-made mass exceeds all living biomass. *Nature 588*, 442-444.

Gray, J. (2021). *Thirteen Paces by Four: Backyard Biophilia and the Emerging Earth Ethic*. London: Dixi Books.

WWF International (2020). *Living Planet Report 2020*. Gland, Switzerland: WWF International.

Chapter 2:
Earth's Animal Nations: It's Time for Human Compassion

Anja Heister

Humanity's moral crisis has been playing out in the catastrophic destruction of nature and the brutal treatment of non-human animals. I see the Covid-19 crisis as a warning to abandon our assumed entitled status as a species (anthropocentrism). This look toward a post-Covid future is toward a world in which we learn to view other animals as equals and we pursue liberty, freedom and happiness together. I also would like to bring some suggestions of how readers can bring more compassion for non-humans into the world.

Human arrogance is unsustainable for the rest of the world

Like so many people around the world, I felt devastated when I first learned about the wildfires in Australia at the beginning of 2020. And when news reported the death toll of one *billion* domestic and wild non-human animals killed in the raging flames in the "Land Down Under," my despair only grew, while my hope for the future went dark. Then Covid-19 hit the planet, sickening and killing humans and non-humans alike, essentially putting humanity on notice that the only certainty is a growing planetary crisis if we do not change our ways of being. As I was reflecting on our planet's predicament, and humans deciding the fate of billions of non-human animals every year, I realised that hope lies in action. Non-human animals depend on us taking action to liberate them from the shackles of human arrogance. In short, humanity needs an intervention.

It is a grim reality that humanity is globally destroying wild animals and wild places at a cataclysmic level. Two recent findings speak volumes. *The 2020 Living Planet Report* commissioned by the World Wildlife Fund for Nature (WWF) documented a 68% decline in vertebrate wildlife populations between 1970 and 2014 globally (1). This report was preceded by another stark warning by the 2019 United Nations-backed landmark report from the Intergovernmental Science-Policy Platform on Biodiversity and Ecosystem Services (IPBES) on the rapid decline of nature, stating that the essential interconnected web of life on Earth is getting smaller and increasingly frayed, and that one million animal and plant species are in imminent danger of extinction, many within the next few decades—all a result of human activity.

Covid-19 and other recent catastrophes are however only the tips of many icebergs hiding mountains of ongoing, unseen human crimes against defenceless non-human animals. Billions of chickens, cows, pigs and other so-called 'food' animals are tortured and killed in the industrialised food system every year, despite the fact that most of us in the Western world have access to a great variety of non-animal and plant-based foods; hundreds of millions of wild animals are trapped and shot by 'sportsmen' for recreation, trophies and commerce all over the world; many more are kidnapped, thrown into cages, and transported to so-called "wet markets" where they await their brutal slaughter. The Covid-19 is only one plague that easily jumps species lines as we penetrate deeper into wild places to take wild animals. The ongoing pandemic holding the entire planet in its viral grip is a prophecy of our future if we do not make radical change in how we live our lives.

As we intensify our exploitation of animals, we have created the tipping point of an imminent mass extinction crisis of plants and non-human animals, which is also nothing less than a moral crisis.

Already in 1967, Lynn White Jr, a medieval historian, warned that an anthropocentric value system lies at the root of our ecological crisis, a view he largely blamed Christianity for promoting. Anthropocentrism is the conviction that moral worth and value lies only in humans and that the non-human world, including non-human animals merely exists to serve human interests. According to this human-entitlement worldview, nature exists to serve human needs and non-human animals—domestic and wild—are merely means to an end. This view has a long history and has been championed by influential historic figures such as the Greek philosopher Aristoteles, who, around 300 B.C. formulated the concept of 'The Great Chain of Being,' (Latin: *scala naturae*), which orders the universe hierarchically with God at the top, progressing downward to angels, humans, animals, plants, and minerals. Others who also declared non-human animals as inferior while elevating humans as superior above them all, included Renée Descartes, the French philosopher and scientist, who in the 17th century conducted horrific experiments on dogs and other animal victims. These atrocities committed by Descartes were possible because he reduced non-human animals to mindless and soulless machines completely devoid of sentience, the capacity to feel pain and other emotions. And while this is clearly an outdated, heartless and false assumption, up to this day, humanity globally operates under the spell of alleged human uniqueness, a view that is normalised by religions, cultures, policies and politics. This artificial separation between humans and the non-human animals—a dualistic view—and the uncompromising self-consumed authority of the former over the latter, is a power structure where humans have come to brutally dominate our fellow animals for our interests. We have been waging a bloody war against other beings, whose tragedy simply is to have been born a non-human animal. Further, our socially constructed separation from other species has not only left us lonely and feeling disconnected to others but is causing our violent and destructive behaviours and attitudes toward nature in general and fellow beings in particular.

Always a ghostly presence, this human entitlement attitude is ubiquitous, culturally embedded, and reflected in all our thoughts, legal frameworks and policies, behaviours, attitudes and actions toward others. Most importantly, it has catastrophic practical implications for non-human animals. We incarcerate billions of non-human animals in tiny cages—sentient cows, pigs, sheep, turkeys and chickens, and treat them horrifically before they are sent to slaughter; the milk a cow mother produces to nurture and raise her child becomes milk for our cereals and coffees and teas; birds' wings, legs from lambs, even ribs from baby pigs and other body parts from non-human animals become our lunches and dinners, no questions asked. We give no thoughts to the fact that our cosmetics, cleaning products and drugs have been tested on an innumerable number of mice, rabbits, dogs and primates who had to endure painful, invasive experiments before being killed. Hundreds of millions of minks, foxes and other 'furbearers' are being forced to endure a miserable, often diseased life in tiny wire cages stacked on top of each other in the tens of thousands in large-scale fur factory operations, before they are electrocuted or gassed to death and their furs are turned into coats and trims for us to wear. The disconnect is so rampant and out of whack that, for instance, at Petsmart stores, dogs' coats sport real fur decoration. Wild animals do not fare any better. Hundreds of millions of them, including deer, antelope, mountain lions, bears, prairie dogs, beavers, elephants, zebras and giraffes are being assaulted, injured and killed by so-called 'sportsmen,' poachers and other human killers for fun, recreation, sport, trophies, commerce and as 'bushmeat.' The destruction of nature, and the brutal treatment of domestic and wild animals are two sides of the same coin—humanity's unwavering belief in its supremacy and right to dominate the non-human world.

Under an anthropocentric value system, other animals are nothing more than objects and human property, things with no more rights than the chair you are sitting on right now, and to whom you have no responsibility to care for their welfare and wellbeing. Under an anthropocentric worldview, the human relationship with the non-human world is dysfunctional, it normalises the exploitation of non-human animals using violence and control. A journalist visiting an industrialised, large-scale chicken 'farm,' stated that, "[T]hese birds are in a condition of pain, lameness, fear, bewilderment and learned helplessness. They are imprisoned in alien, dysfunctional bodies, in confinement facilities full of filth including atmospheric poisons and contaminants. What will it take for our species to realise that treating creatures this way is a crime, an iniquity?" (2).

According to feminist writer Carol Adams, the most efficient way to ensure that humans do not care about the lives of animals is to transform living and feeling non-human subjects into non-human *objects*. When subjects are turned into objects, we suspend our emotions and we no longer care. The common view here is that they (the animals) are not like us (humans)—*they are just animals*. In other words, they are inferior and as such, we believe that neither their lives nor their deaths matter to them—that's what they are here for. We are not required to feel empathy for things when we merely consume a food product—*it's just food*. We have forgotten that no pig dies as pork, no elk dies as game, no bobcat dies as fur and no pangolin dies as an expensive delicacy. Animals die as individual beings, and how they die "matters a great deal to the one who is dying" (Adams 2007).

"Nature is sending us a message ..." (Carrington 2020) warned the United Nations' environmental chief in March 2020 as the significance and fatal dimensions of the Covid-19 virus outbreak started to sink into the collective consciousness of the global human population. Are we listening? Are we asking what other animals want from us? How will we respond?

We live on a planet to which no one species has exclusive rights. As one of approximately 8.7 million species (6.5 million species on land and 2.2 million in oceans), according to one estimate, humanity has emerged and evolved, amidst the companionship of non-human animals, with whom we share genes, cells, organs, bodies, behaviours and minds—and, which today, identify us as the human species that shares an evolutionary track with other species.

We are surrounded by other beings with their own minds and sentience, the ability to think, to experience emotions such as happiness, joy, gratitude, grief, pain and suffering. The fact that other animals share capabilities to experience emotions and cognition and share consciousness has been increasingly embraced (thus proving Descartes wrong, and supporting common sense) ever since on July

2012, a group of established scientists published the "Cambridge declaration of consciousness," which essentially concluded that all mammals, birds and other species have the neuronal architecture that underpins consciousness. The declaration, which was signed in the presence of the late, renowned physicist Stephen Hawkins, emphasised that "non-human animals have the neuroanatomical, neurochemical, and neurophysiological substrates of conscious states along with the capacity to exhibit intentional behaviours." Consequently, the weight of evidence indicates that humans are not unique in possessing the neurological substrates that generate consciousness. Non-human animals, including all mammals and birds, and many other creatures, including octopuses, also possess these neurological substrates (Panksepp et al., 2012). The declaration essentially consolidated contemporary research findings, some of which corrected previously assumed hypotheses in science that leave the assumption of human uniqueness in the rear mirror. By doing so, the declaration corrected previously held beliefs about the hierarchical thinking of evolution that views animals' merely being less-than-humans, and noted instead that the behaviour, neurophysiology, and neuroanatomy of birds and mammals are a striking case of *parallel evolution* of human consciousness. In sum the declaration, along with other research, moves animal experiences out of the shadow of human uniqueness and positions them on par with *Homo sapiens.* And not only that, but scientific findings confirming non-human and human comparability open the door to considerations that other animals may by far surpass whatever humans generally possess. For example, other species have sensory-perceptual capacities such as olfaction in dogs, audition in cetaceans, or use of ultrasound by bats that are far more sophisticated than what humans possess.

Case in point, marine scientist Lori Marino has stated, that "[C]learly, echolocation—the use of high frequency sound for processing information—by cetaceans is more than a fancy receptor device [usually this ability is described in non-cognitive terms]. It is a highly sophisticated cognitive capacity that humans lack—and all that that implies" (Marino, 2010). In the same vein, Marc Bekoff, professor emeritus of Ecology and Evolutionary Biology, has said that the more we study animals, in a non-invasive way of course, the more we learn about their rich emotional lives and cognitive skills making up their amazing world. Indeed, chickens are courageous and empathetic to others' distress, so are mice; foxes use the Earth's magnetic field to measure distance and direction when they pursue prey under the snow; at one time, a male wolverine (the largest member of the weasel family) in the Canadian mountains, a species long thought to be solitary and aggressive, was observed visiting and spending time playing with his children living at their mother's den. Wildlife cameras have captured coyotes and badgers having fun traveling together to hunt mice and voles as partners.

The fact that a growing number of humans know that other animals have minds and emotions and sophisticated social relationships with members of their own and other species within their homelands - ecosystems and habitats - has consequences. Thus, implicit in the Covid -19 crisis is an urgent call to come out of the closet, so to speak, and at the risk of being belittled speak up about our views and demand a change in human treatment of non-human animals. Non-human animals belong here—they are Earth's Animal Nations—they are not our property, nor are they things, objects or resources for humans to use. Non-human animals are here on this planet for themselves, and their right of existence is independent from whether or not humans assign any (artificial) value to them. They are essential to the great web of life. And under the threats of climate change, Covid-19, there can no longer be "us versus them" but rather, we are all in this together with the effects of the crises, current and future ones, that affect all of us.

We now have an opportunity to pause, to take a moment of reflection on what needs to be changed and how to implement that change.

We need to transform ourselves from 'superior takers' to 'humble and compassionate helpers' while recognising that non-human animals too are holders of property rights and bestowed with desires shared with humans-- liberty, freedom and happiness. Just like Molly B, a cow, who escaped a slaughterhouse, jumped fences and ran as fast and far away as she could. Luckily, compassionate humans took her to safety--a farm animal sanctuary where she got to live out her life among other rescued cows, pigs, sheep and turkeys (Kuglin 2014).

The animal rights journey began with Peter Singer and Tom Regan's "moral extensionism" approach in 1975 and 1983 respectively. Singer's influential book *Animal Liberation* certainly galvanised the animal advocacy movement. In revisiting Jeremy Bentham's quote, "The question is not, Can they reason? nor, Can they talk? but, Can they suffer?" Singer argued that a utilitarian calculation—achieving the greatest good for the greatest number, for example achieving the greatest happiness or the greatest satisfaction of interests, must include all who can be made happy, all who have interests. Thus, sentient non-human animals must be included in all our moral considerations. Singer also introduced the term 'speciesism' (coined by Richard Ryder in 1970) into the mainstream of the animal advocacy movement. When we treat members of our own species favourably simply because they are humans, while mistreating other animals simply because they belong to a different species while sharing morally relevant similarities such as feeling pain, this is called speciesism. This concept is closely associated with our notorious human-entitlement attitude, and it is also related to racism and sexism, all of which are characterised through one group asserting domination, power and control over another perceived as inferior. Rejecting utilitarianism altogether, Tom Reagan's *The Case for Animal Rights* in 1983 argued that most mammals aged one year or older are "subjects-of-a-life" whose interests and needs had to be considered. Both philosophers challenged the anthropocentric ethics of utilitarianism, the Western tradition that grants moral standing exclusively to human beings and considers the rest of nature, including non-human animals, as sources of utility, or mere means for human ends. This was followed by a wave of feminist writers, who argued against the "like us" approach put forward by Singer and Reagan, and made the case for the importance of positive emotions, feelings and care in our relationships with other beings.

Anthropocentrism is incompatible with emotions we ground our own, human-to-human relationships in such as love, awe, respect and empathy and compassion. It undermines these crucial, pro-social emotions because under a human-entitlement worldview, we only value others so far as they serve our interests, we do not value them for themselves because it is all about us. On the other hand, love and other positive emotions go beyond ourselves and it does not care how different the other being is. For example, "the love involved in friendship is an *other*-centered emotion," (McShane 2007). In other words, I do not love my friend just because the relationship serves me well... she is always there for me when I need her... instead I also love my friend for the person she is and I see great value in her even if she were not my friend... independent from my own interests, and no matter how different she is from me.

So, we need to include emotions in our relationship with other animals to benefit them. We need to transition from a "it's all about us" attitude to a "we are all in this together" goal. We can no longer deny other animals their rights to live without being harmed by humans. Perhaps most importantly, from now on, other animals and their homelands need to be part of our thinking, planning, values, religion, culture and politics, and the choices we make—always, with the question in mind, "how is this going to affect our fellow earthlings?" and, "what can I do to bring more compassion into this world to support other animals?" And, "how do we work together?"

The wellbeing of the Earth as a whole, and the wellbeing of its non-human and human members are connected and interdependent. However, we cannot forget that "after all, the Earth and most of its members can get along very well without us, but we cannot live without them" (Spencer 1996, p. 9). Thus, it is incumbent upon us to live in a relationship with other species that is grounded in the desire to see all of us flourish and follows an ethic of respect that does not depend on other animals bearing the burden and paying the costs for our own perceived benefit. Therefore, humans must strive for acknowledging the connection and similarities with other species while appreciating, learning from, and celebrating differences. And not only that, we need to go further, and speak and stand up for non-human animals and their interests, and include their rights to wellbeing and land into our countries, provinces and states' constitutions, which currently only address and protect human interests. However, social change is essential first, in order to reform the legal framework and to free non-humans from the property shackles of human arrogance and violence.

We now must take action to make concepts a reality that would elevate the status and our treatment of non-human animals as co-citizens, including their right to habitat (Donaldson, 2020) and true sanctuary for the billions of 'food' animals and non-human animals used in experiments, where they can live out their lives under compassionate human care.

The late theologian Thomas Berry's approach in our goal to overcome our ingrained sense of superiority is clear in his proposal that "the Earth is a communion of subjects, and that rights originate where the universe originates and not from human jurisprudence" (Cullinan, 2002: 108). This means "we cannot claim that humans have human rights without conceding that other members of the Earth Community also have rights" (Cullinan, 2003: 108). For this to happen, nature and its wild and domestic non-human animals need to be released from their legally enshrined property status. Instead, the more-than-human world must be recognized as having rights to exist, persist and flourish, with people having a moral obligation and authority to enforce nature's rights on behalf of ecologies and their denizens. This enormous transformation of our relationship with nature has been taken up by the rights of nature movement (Sólon, 2018) and associated legal initiatives (e.g. Earth Law Center, 2019), which provide us with a much-needed holistic ethical and legal framework that re-embeds humans into the ecological context and gives nature a voice. This spiritual and practical, justice-based vision of Earth democracy has already begun to shape a crucial egalitarian relationship with the more-than-human world. For example, in 2008 Ecuador included rights of nature in its new Constitution and, more recently, the Maori tribe in New Zealand achieved the legal recognition of a large river as an ancestor with legal personhood. Efforts by the organisations including the USA-based Non-human Rights Project and People for the Ethical Treatment of Animals (PETA), the Netherlands-based Party for Animals that is represented in the House of Representatives, the Senate and even the European Parliament, all work toward sustainability, social issues and strengthen the recognition and protection of the fundamental rights of non-human animals. The American writer Henry Beston wrote in 1928, "For the animal shall not be measured by man [and woman]. In a world older and more complete than ours, they move finished and complete, gifted with the extension of the senses we have lost or never attained, living by voices we shall never hear. They are not brethren [or sisters], they are not underlings: they are other nations, caught with ourselves in the net of life and time, fellow prisoners of the splendour and travail of the earth." Critically, their wellbeing and survival depends on us and we need to do everything we can to protect and save them, for themselves and for our own sanity.

What you can do

• One of the most effective ways to make a profound difference for all non-human animals, domestic and wild, and for the health of the planet more broadly, is by switching to a plant-based vegan diet.

• Start your own organisation to fight for the rights of non-human animals and against corporate enslavement of our fellow earthlings.

• You can affect change by getting involved in organisations that work toward the advancement of animal rights and the protection of individual wild animals.

• You can get involved in your local, regional or national legislature, where all too often bills detrimental to wild and domestic animals are being passed. Testify at hearings, call your representatives and educate them.

• Check every product before purchasing it to make sure it was not tested on non-humans or otherwise contributed to their exploitation.

• Consider running for office. This is the strongest thing you can do to promote non-human and Earth-friendly policies and vote for legislators who champion these. Bring attention to the plight of non-human animals by writing letters to your local newspaper. If you belong to a congregation, the peace movement, or any environmental, social justice, political or conservation organisation, question them on their stance on hunting, fishing and trapping (you'll be surprised), as well as a plant-based diet.

• Visit animal sanctuaries and get to know their animal residents to learn more about their individual personalities, interests and needs.

• Speak out! Contact your elected officials about the effect of any bills on animals.

References

Adams, C. (2007). Introduction. In: Donovan, J. & Adams, C., eds. *The feminist care tradition in animal ethics: A reader.* New York, NY: Columbia University Press, p. 26.

Bekoff, M. (2011). *Empathic chickens and cooperative elephants: Emotional intelligence expands its range again.* Available at https://www.psychologytoday.com/us/blog/animal-emotions/201103/empathic-chickens-and-cooperative-elephants-emotional-intelligence (Accessed June 2020).

Beston, H. (1928). *The Outermost House: A Year of Life on the Great Beach of Cape Cod.* Available at https://www.yumpu.com/en/document/view/63985508/download-the-outermost-house-a-year-of-life-on-the-great-beach-of-cape-cod-book-pdf-epub (Accessed June 2020).

Carrington, D. (2020) Coronavirus: 'Nature is sending us a message', says UN environment chief. The Guardian, 25 March. Available at https://www.theguardian.com/world/2020/mar/25/coronavirus-nature-is-sending-us-a-message-says-un-environment-chief (Accessed June 2020).

Coyote and Badger Playing Together - *California Wildlife Camera Footage (2020).* Available at https://www.youtube.com/watch?v=2bICTWNRrGE (Accessed June 2020).

Cullinan, C. (2003). *Wild Law: A manifesto for earth justice.* Totnes: Green Books.

Davis, K. (2020). *An Open Letter to The Guardian.* The Guardian, 20 August. Available at https://www.upc-online.org/thinking/200820_an_open_letter_to_the_guardian.html (Accessed August 2020).

Donaldson, S. (2020). *Animals and Citizenship.* Available at https://www.humansandnature.org/animals-and-citizenship (Accessed June 2020).

Earth Law Center (2019) *What is Earth Law?* Available at https://is.gd/IXKtw1 (Accessed July 2019).

IPBES (2019) *Global Assessment Report on Biodiversity and Ecosystem Services.* Intergovernmental Science-Policy Platform on Biodiversity and Ecosystem Services, Bonn.

Kuglin, T. (2014). *Famous cow who escaped slaughter moves to Helena Valley.* Available at https://helenair.com/news/local/famous-cow-who-escaped-slaughter-moves-to-helena-valley/article_623d804b-8560-58a8-9dce-67e82fb8dd68.html (Accessed September 2020).

Marino, L. (2010). *Trans-species perspective. On the human.* Available at http://onthehuman.org/2010/11/trans-species-perspective/ (Accessed June 2020).

McShane, K. (2007). Anthropocentrism vs. nonanthropocentrism: why should we care? *Environmental Values 16* (2).

Panksepp, J., Reiss, D., Edelman, D., et al. (2012). T*he Cambridge declaration on consciousness.* Available at: http://fcmconference.org/img/CambridgeDeclarationOnConsciousness.pdf (Accessed July 2020).

Regan, T. (1985). *The Case for animal rights.* Berkeley and Los Angeles, CA: University of California Press.

Singer, P. (1975). *Animal liberation: A new ethics for our treatment of animals.* New York: Avon Books.

Sólon, P. (2018). *The rights of mother earth.* In: Satgar V, ed. *The Climate Crisis: South African and global democratic eco-socialist alternatives.* Wits University Press, Johannesburg, p. 107–30.

Spencer, D. (1996). *Gay and gaia: Ethics, ecology, and the erotic.* The Pilgrim Press, Cleveland, Ohio.

White, Lynn Jr. (1967). The history of our ecological crisis. *Science 55,* p. 3767.

WWF International (2020) *Living Planet Report.,* Gland.

Online resources

1. Living Planet Report, (2020). https://livingplanet.panda.org/en-US/what-is-the-living-planet-index (Accessed December 2020).

2. Davis, K. (2020). An Open Letter to The Guardian. *The Guardian,* 20 August. Available at https://www.upc-online.org/thinking/200820_an_open_letter_to_the_guardian.html (Accessed August 2020).

Chapter 3:

Rewilding on a Global Scale: a Crucial Element in Addressing the Biodiversity Crisis

George Wuerthner

We are entering what some have termed the Anthropocene Epoch, where some suggest humans are now like a geological force. We are now experiencing the Sixth Great Extinction. The current extinction rate is nearly 1,000 times higher than during the pre-human era, and traditional conservation movements will not work fast enough to save the natural world.

Our growing population now surpasses the ability of the Earth to regenerate the basic needs of society. Indeed, humans and our livestock now comprise 96% of the biomass on Planet Earth. All the other animals, from shrews to elephants, only make up 4% of Earth's living life.

A human presence overwhelmingly influences the planet. The human footprint already uses more resources and exceeds Earth's capacity to absorb our wastes (like carbon). One estimate even suggests that even our present population requires 1.6 Earths (1).

With the global population growing, we may need two planets the size of Earth to supply our needs. We are exceeding carrying capacity.

The numbers are sobering. The Intergovernmental Science-Policy Platform on Biodiversity and Ecosystem Services (IPBES) predicts that a million species will go extinct in the coming decades.

According to the report, the average abundance of native species has declined by 20% since 1900. Other groups that have suffered significant declines, including more than 40% of amphibian species. At least 680 vertebrate species had been driven to extinction since the 16th century (2).

Other worrisome conclusions are that human actions have significantly altered three-quarters of the land-based environment and about 66% of the marine environment. More than a third of the world's land surface and nearly 75% of freshwater resources are now devoted to crop or livestock production.

Add to the human modification of land and water, is on Green House Emissions (GHG). Due to human-caused C02 rise, we see more floods, stronger hurricanes, more wildfires, significant droughts, and rising seas, not to mention species decline (coral reefs) as a consequence of climate change.

Agriculture is responsible for significant GHG emissions, and also one of the most destructive of human activities. Typically, agriculture favours one or a few plants or animals over large Earth areas. Since there is only so much soil, water, and land, this naturally reduces the carrying capacity for native species. A change in diet is probably what most Western countries could do immediately to reduce carbon emissions (3). Fruits and vegetation production has the lowest contribution to GHG emissions, while meat and dairy have the highest emissions (4).

A 212-page online report published by the United Nations Food and Agriculture Organization says 26% of the Earth's terrestrial surface is used for livestock grazing. Livestock feed crop cultivation occupies one-third of the planet's arable land (5).

All of these issues are ultimately related to the elephant in the room: human population growth. The world's population is estimated to grow from its current 7.5 billion to over 11 billion by 2100. Even if every person used fewer resources per capita, the human population's sheer growth would guarantee biodiversity losses (Cafaro and Crist 2012).

Rising competition for resources ensures conflicts and increasing poverty. Desperate people are more likely to ignore any constraints to protect non-human species. So increased attention must be given to helping people adjust their fecundity through education of women, contraception, counselling, and changing men's attitudes about masculinity.

The opportunity

The strategy to counter species endangerment and ecosystem collapse is to protect natural landscapes.

Parks, preserves, and other strict conservation areas fare the best in protecting species, though even in such preserves, there is a decline - typically because of inadequate funding, small size, and other well-documented reasons (Wuerthner et al. 2015)

However, many anthropology and other academic pursuits characterise parks, reserves, and other protected areas as "fortress preserves" because they prohibit human resource uses.

Those advocating this position suggest that wilderness is only a "cultural construct" primarily of elitist white males. Advocates of the Anthropocene perspective argue that there is no "pristine" wilderness.

However, that is a straw man since most informed parks, wilderness, and preserve advocates recognise that humans have been influencing the natural world for thousands of years. Our homo sapiens ancestors likely killed off the Neanderthals and other primates. There is abundant evidence that Indigenous Peoples from Australia to Hawaii to North and South America, once they arrived in lands without people, were responsible for the extinction of many species.

Nevertheless, even in a world of global human influences like climate change, there are real social impact differences. New York City is almost entirely a human construct, while the Arctic Wildlife Refuge in Alaska is primarily under natural processes and influence. Wilderness or wildlands means that lands are self-willed. Places like the Arctic Wildlife Refuge can reasonably be termed "natural" and "wild." And these self-willed lands are critical to stemming the loss of biodiversity (Wuerthner et al. 2014).

Anthropocene advocates and critical social scientists are often indifferent to the loss of wildlands and/or biodiversity or the domestication of the Earth.

To the degree they care about biodiversity loss, it is typically entangled in human utility. Ironically, this attitude is nearly identical from both ends of the right or left political spectrum.

Most of those advocates believe that humans are a part of "nature" therefore, there is no distinction between nature and the human world. Indeed, it is argued that since humans have lived in just about every landscape (except Antarctica), there is no such thing as "wild nature" or the absence of human influence.

Yet if everything humans do is natural, restricting destructive practices like logging, road construction, overhunting, overfishing, overgrazing, soil erosion from agriculture, even air and water pollution becomes impossible to justify.

Critics characterise such "fortress preserves" as a form of "ecofascism" or 'imperialism." Ironically, such social critics and Anthropocene boosters implicitly suggest that humans are masters of nature - which in effect advocates human "supremacy" which is a form of nature imperialism.

However, many Anthropocene boosters and social scientists argue traditional conservation efforts ignore the poor or disenfranchised humanity. This was particularly true in the past.

Most nature preserves are in places with lower human population density and development where the fewest people are impacted. Social justice advocates suggest that such preserves in these areas put wildlife ahead of humans, and some argue they are essentially undemocratic.

Social justice advocates argue that putting resource use limits on any people, particularly poor or indigenous people, is an unfair burden. Indeed, in all instances, parks and wilderness advocates must make serious attempts to compensate people for any costs of conservation.

Nevertheless, the assumption that continued resource exploitation is the only option for attaining social justice needs to be examined.

One advantage of "fortress preserves" is that they imply limits.

One of the philosophical foundations of park, wilderness, and other designations is restraint. It is an acknowledgment that humans should embrace constraints.

Furthermore, parks, wilderness, and wildlife preserves often benefit local people, because such parks and preserves provide human services such as clean water and air, more stable landscapes (fewer floods and severe drought), and durable natural ecosystems. In some cases, economic opportunities exist in eco-tourism and even existing resource exploitation. For instance, marine reserves often increase the fish available to human consumption outside of preserves.

The argument about whether parks and preserves harm poor people often focuses on who has the "right" to exploit natural systems, not whether anyone should exploit them. While "rights" are argued, responsibilities to the rest of the natural world are often ignored.

Many people support "sustainable development," but what they discuss is "sustainable economic development," not sustainable ecosystems. In other words, a forestry company might cut fewer trees so that there is a constant supply of trees for the mill, but logging at this level could still degrade wildlife habitat. Such an approach might maintain the timber industry, but it may not sustain the forest ecosystem.

At least a quarter of the global land area is traditionally owned, managed, used, or occupied by Indigenous Peoples. Indigenous lands, while faring better than unprotected landscapes, still exhibit a decline in species (6). Nevertheless, in many cases, Indigenous Peoples are increasingly part of the global economic system and can have a significant impact on biodiversity as well.

With modern technology impacts from even Indigenous Peoples can be substantial. Many Indigenous preserves still permit things like hunting, slash and burn agriculture, logging, grazing, and other resource exploitation. These resource extraction uses can be absorbed without damaging ecosystem sustainability if done over limited areas. Nevertheless, they are not a replacement for the traditional conservation approach of preserves with limited human impacts.

Therefore, wildlands protection is even more critical because they effectively slow and sometimes reverse biodiversity losses.

Advocates of parks and preserves argue for ethical and moral reasons to support an ecocentric perspective, which suggests species should be protected and preserved regardless of their perceived value to human interests. Such a view argues that it is morally wrong for humans to cause the extinction of other species.

The traditional approach of good-sized parks and preserves where there is a minimum of human impact has been shown to be the most sustainable for biodiversity.

Some biologists like E.O. Wilson, suggest that we should protect 50% of the land for nature. In his book *Half-Earth,* Wilson calls for devoting half of the surface of the Earth to save or at least slow the loss of species.

The solution

The first consideration to review is what conservation biology principles, often called Island Biogeography, tell us about conservation protection. As a rule, the larger the protected area or "island of habitat," the more likely it will be to preserve the home for many species. A second tenet is to have linkages or corridors between the larger parcels, so there is some movement of plants and animals between the larger properties.

There are still large areas of the Earth where the human population is relatively light, and the amount of land that is mostly "self-willed" is significant. For example, much of Canada's Boreal Forest is still intact—although under assault by logging and energy development. The creation of extensive new parks preserves and wilderness on these lands is still an option, especially if there is support from local people.

While no doubt building a global conservation network will occasionally entail displacement of humans just as we displace people to make a highway or reservoir, there are other ways to achieve some of the Half-Earth goals that do not require the involuntary removal of people.

One useful measure is to eliminate or reduce resource extraction in areas where productivity is low. In these areas, it requires many acres of land to produce the same product in more fecund areas. Resource exploitation is often only economical by ignoring the real ecological damage from development. For example, over much of the Great Plains, soil erosion due to farming is excessive; however, soil loss is not included in the cost of the crops grown on these lands. If these costs were internalised, much of the more marginal land uses would no longer be suitable for exploitation.

Plus, the ecological damage from resource use is often more significant in such areas.

For instance, in the Nevada desert, a cow may require up to 250 acres a year to sustain itself, while the same cow can be grown on an acre or two in a moist, warm region like Georgia or Alabama.

Water is scarce in deserts. The water influenced areas known as riparian areas are the focal point for 70-80% of all wildlife in desert areas. But these are the same areas that cows congregate to obtain water, green vegetation, and shade, often to the detriment of the riparian areas and watershed. Thus, livestock utilisation of these areas by livestock can have a disproportionate impact on biodiversity.

It would be more effective to grow a cow in Georgia, where there would be fewer biodiversity losses.

There is a passive recovery of natural areas when inefficient land uses are abandoned. For instance, by the late 1800s, 85% of Vermont's forests were cut over for farming and lumber. But there were better lands for both growing livestock and crops and trees in other parts of the country, and over time, the farms and timber operations were abandoned, and the forests came back on their own. Today 80% of Vermont is forested.

The subsequent reforestation led to the recovery of many wildlife species extirpated due to either overhunting and/or habitat loss. Wildlife, including moose, marten, fisher, and even lynx, once gone from Vermont woodlands, are now relatively common.

A similar passive restoration is also possible over much of the Great Plains. For decades, people have been fleeing the northern Great Plains, and most of the counties in this region of Montana,

North Dakota, South Dakota, and Wyoming have far fewer people today than in the early 1900s. Many of these counties now qualify as "Frontier Counties" by definition of the 1890s census, which suggested any area with fewer than 2 people per square mile was "frontier."

With the decline of the human population, there is an opportunity for significant natural ecosystem recovery. Efforts like the American Prairie Reserve, which has acquired lands in eastern Montana and restored bison, prairie dogs, and swift foxes, is an excellent example of the opportunities that will increasingly occur in this region.

Given that much of the Northern and Central Great Plains is used to grow feed for livestock such as feeder corn or soybeans, another way to obtain more land for nature preserves is to eliminate or at least reduce meat and dairy from our diet.

Across the US, some 130 million acres are in hay production and/or pasture for livestock, 90 million acres more or less is used to grow feeder corn, and another 75 million acres are used for soybeans. To put this into perspective, Montana is 93 million acres in size, so this is equivalent to three times the acreage of Montana.

These figures are gross amounts since some of the corn, soy, and other crops may be used for direct human consumption or other products like ethanol fuel. Nevertheless, the bulk of these acres are growing crops to feed livestock. Thus, a reduction in meat and dairy consumptions offers an incredible opportunity to restore many parts of the American landscape. With this, and additional points that I make above, similar considerations apply to many other parts of the world.

A transition to renewable energy is also a necessary component of any solution. As previously mentioned, fossil fuels are among the leading sources of GHG emissions, not to mention the destruction and fragmentation of wildlife habitat because of energy development related roads, drill pads, and native vegetation clearing.

We cannot reasonably slow or reverse climate warming without a global commitment to renewable energy. Fortunately, there are numerous opportunities for reducing fossil fuel consumption, including more efficient transportation, better insulation of buildings, and expansion of geo-thermal energy, wind energy, and solar energy—all of which are becoming more efficient and effective (Butler and Wuerthner, 2012).

Conclusions

While current human population growth, along with rising GHG emissions, poses the most long-term threat to both wildlands and biodiversity, there are also opportunities to strive towards the goal of protecting half of the Earth for biodiversity and other creatures. Not only is this goal achievable, but it is essential if human life is to be preserved at more than a survival existence. Furthermore, protecting half of the Earth is also a moral and ethical obligation of human society.

References

Butler, T. & Wuerthner, G. (2012). *Energy—Overdevelopment and the Delusion of Endless Growth.*

Cafaro, P. & Crist, E. (2012). *Environmentalists Confront Overpopulation.*

Wuerthner, G. et al., (2014). *Keeping the Wild—Against the Domestication of the Earth.*

Wuerthner, G. et al., (2015). *Protecting the Wild—Parks, and Wilderness the Foundation for Conservation.*

Online resources

1. https://www.worldwildlife.org/threats/the-human-footprint (Accessed November 2020).
2. https://www.un.org/sustainabledevelopment/blog/2019/05/nature-decline-unprecedented-report/ (Accessed November 2020).
3. https://ourworldindata.org/food-ghg-emissions (Accessed November 2020).
4. https://climatecommunication.yale.edu/publications/climate-change-and-the-american-diet/2/ (Accessed November 2020).

5. https://www.smithsonianmag.com/travel/is-the-livestock-industry-destroying-the-planet-11308007/#:~:text=A%20212%2Dpage%20online%20report,by%20livestock%20feed%20crop%20cultivation (Accessed November 2020).

6. https://www.un.org/sustainabledevelopment/blog/2019/05/nature-decline-unprecedented-report/ (Accessed at November 2020).

Chapter 4:

Case Study 1: A Peace Park Founded by the Indigenous People of Kawthoolei

Karen Environmental and Social Action Network

Originally published by *Radical Ecological Democracy* and then made available through a Creative Commons BY 4.0 licence (https://creativecommons.org/licenses/by/4.0/) as: Karen Environmental and Social Action Network (KESAN) (2020) Salween Peace Park: for all living things. *The Ecologist*, 8 September. Available at: https://theecologist.org/2020/sep/08/salween-peace-park-place-all-living-things (accessed February 2021). The only changes made were to bring the formatting style in line with that of this book. The piece has been given a new title here.

Editor's note: It is a real pleasure to be able to include this case study in the book, through a Creative Commons licence. Not only does it showcase Indigenous thinking on the crucial issue of protected land but the ideas presented also tie in beautifully with numerous themes covered elsewhere in the book. My thanks go to the Karen Environmental and Social Action Network for writing such a fascinating and important piece.

The Karen People have lived in our forest home for 2,758 years according to our calendar. Our lands and waters play many important roles in everyday life and in our future prosperity. They are core to the subsistence practices of our communities.

Karen territories boast fertile soil, where the 'Ku' shifting cultivation system is used to grow vegetables and other foods rotationally, allowing nature to recover. The rivers of our Karen territories, including the Salween, provide a means of reliable transport and trade, as well as a rich source of fish. Our people forage for wild foods like bamboo shoots, banana fruits and flowers, honey, mushrooms, and edible ferns in verdant forests.

We peacefully coexist with rare and endangered animals like the Sun Bear. Our communities gather forest materials to build and maintain homes, to make various tools and create art.

Kawthoolei

Our ancestral territories are a repository for our history, culture, and beliefs. Karen communities are predominantly animist, and our practices and culture are deeply intertwined with and situated within our ancestral territories, which we call Kawthoolei.

For our communities, the conservation of nature is vital to the conservation of our own culture. The health of one directly corresponds to the health and prosperity of the other. This is expressed through cultural traditions and taboos that encourage sustainable use of some resources and forbid the harvesting or use of others. They are observed seriously.

We are the best custodians of our ancestral territories. This is demonstrated by the rich biodiversity of Kawthoolei, which is situated in the Indo-Burma biodiversity hotspot and is of global significance to nature conservation.

Many areas in Myanmar have been deforested, with animal habitats destroyed, and plant species lost, but in our Karen homeland healthy populations of threatened and near threatened wildlife can thrive.

Mines

For decades our culture and Kawthoolei homeland have been under assault. The conflict in this region is one of the longest-running civil wars in the world. Since 1949, a year after Myanmar gained independence from Britain, the Karen have been fighting for political independence from Myanmar.

In over 70 years of armed conflict, many thousands of Karen people have experienced genocide, torture, and sexual violence at the hands of the armed forces of Myanmar. Hundreds of thousands of civilians have been displaced throughout the course of the conflict, with many fleeing to Thailand or becoming internally displaced peoples.

More recently, the main challenges that communities in Kawthoolei face derive from logging, mining, infrastructure projects including road and bridge construction, and a series of government-proposed mega-hydropower dams on the Salween River. Communities also face threats from private agribusiness, logging and mining concessions, which are predominantly granted to outside interests and conducted within Karen ancestral territories without communities' permission or any form of compensation.

The Myanmar Government's constitution claims all land, waters, and natural resources for the government. Created without the involvement or consent of indigenous communities, these laws do not recognise the tenure rights and cultural practices of the Karen people. They seek to evict Karen communities from ancestral territories and eradicate traditional forms of 'Ku' shifting cultivation.

The Myanmar Government's push for territorial domination and the monetisation of natural resources and land, conducted through military violence, is destructive to the Karen Peoples and our lands.

Mining activities continuously create challenges for our culture, livelihoods and traditional forms of conservation. Common methods of gold mining are disruptive to local wildlife, destroying habitats and poisoning water sources with mercury and engine oil.

In parts of Kawthoolei, the sheer amount of soil and silt that have to be moved to access the subterranean gold has led to river sedimentation, reduced access to clean water for drinking and bathing, and damaged aquatic ecosystems. Resultant chemical runoffs and air pollution have also caused health issues, including skin and respiratory problems. Yet gold and stone mining continue and recorded mining activities have increased since the 2012 ceasefire.

Resistance

Karen communities have resisted these externally imposed destructive development projects since they began during the colonial era. During the time of British colonial control, communities worked together to protest British logging concessions in Karen ancestral territories, and negotiated with local British administrative officers to resolve disputes between them and communities.

Since Burma's independence, Karen communities have continuously called for the recognition of their rights to their ancestral territories and a peaceful and stable life. This has primarily taken the form of public protests, and the creation and dissemination of reports, documentaries, and songs about the issues of destructive development and impacts of armed conflict that we face.

In recent decades these protests have been primarily focused on the continued presence of the Burmese army in Karen territories, and the threat this poses to communities' lives and livelihoods. Protests have also resisted Thai and Chinese-backed mega hydropower dams proposed to be built on the Salween River.

Karen communities have used a broad variety of strategies to resist unwanted and destructive activities in Karen areas. Public protests and marches are conducted annually on March 14th, the International Day of Action for Rivers and Against Dams, while smaller protests are conducted throughout the year against specific proposed development and/or investment projects, and the increased militarisation of the area by the Myanmar Army.

Communities also resist through celebration, gathering together to promote and strengthen Karen culture and history on important days throughout the year including August 9th, World Indigenous Peoples' Day, January 31st, Karen Revolution Day, and Karen New

Year which is on a different day each year based on the Karen calendar.

Our most tangible success from these decades of resistance is the declaration of the Salween Peace Park (SPP) in December 2018, and subsequent election of its General Assembly and Governing Committee in April 2019.

Hope

The Salween Peace Park is a grassroots, people-centred alternative to the Myanmar government and foreign companies' plans for destructive development in the Salween River basin.

Instead of massive dams on the Salween River, we propose small hydropower and decentralised solar power. Instead of large-scale mining and rubber plantations, we call for eco-tourism, sustainable forest management, agroforestry and organic farming. Instead of mega projects that create conflicts and threaten the resumption of war, we seek a lasting peace and a thriving ecosystem where people live in harmony with nature.

The Salween Peace Park empowers our indigenous Karen communities to guide local development and conservation in line with traditional knowledge and cultural practices. By basing local governance in the hands of the community, the Salween Peace Park enables the conservation of nature and Karen culture, and the pursuit of a peaceful and stable life for local communities, something that is denied to them by the Myanmar government's laws and military ambitions.

In bringing the many efforts of communities across the area together into a coordinated unit, the SPP also seeks to upscale and strengthen the voices and aims of its communities. The SPP General Assembly is comprised of representatives from individual communities and Karen governing bodies, allowing communities to voice their opinions and concerns, and support their neighbours to present a strong united front in opposition to the destructive development and militarisation that threaten their everyday lives.

Local governance structures have been established, with

power stemming from the grassroots-level upwards, and a charter representing the principles laid out by the SPP's communities has already been developed. Members of the General Assembly are now working with knowledgeable community members in a series of working groups to strengthen the governing body and develop a series of initiatives to improve the lives of Karen communities inside the SPP. A master plan is also being developed, guided by local communities, to build a roadmap towards achieving their aspirations of peace and self-determination, environmental integrity, and cultural survival.

The new Myanmar government has promised to lead the country toward a devolved, federal democracy, but, so far they have not delivered. The Karen are not waiting idly for this: the Salween Peace Park is federal democracy in action.

The authors

This article was written collectively by members of the Karen Environmental and Social Action Network (KESAN). KESAN is a community-based, non-governmental, non-profit organisation that works to improve livelihood security and to gain respect for indigenous people's knowledge and rights in Karen State of Burma, where the violence and inequities of more than 60 years of civil war have created one of the most impoverished regions in the world.

Chapter 5:

Fostering a Love of the Living World – or, The Need for a Grand Revival of Natural History

Joe Gray and Reed Noss

[T]his age of biological decline is, not coincidentally, also an age of human indifference to the more-than-human world. Wild nature has been replaced by human-dominated landscapes, circumscribed by patterns and processes never before seen, the consequences of which have been to insulate humanity from other species and wilder landscapes. We now live in a world where it matters more whether it is Friday or Saturday than if it is autumn or winter [...] The loss of biological diversity can only be recognized by someone as a crisis if she has a relationship with nature that is personal and immediate.

— Stephen Trombulak and Tom Fleischner (2007)

"Fake animal news abounds on social media as coronavirus upends life." This was the headline of a piece written for the *National Geographic* website by Natasha Daly in March 2020. In the article, Daly collated examples of news stories that reported on other-than-human animals thriving and revelling in a world temporarily abandoned by humans – items that went viral before subsequently being debunked (Daly, 2020). Swans and dolphins, for instance, had supposedly returned to Venice's deserted canals, while a group of elephants had reportedly ambled into a village in China, drank corn wine to the point of inebriation, and then fallen asleep in a nearby tea field. There were problems, Daly observed, with each of these stories.

The swans were regularly seen, before the Covid-19 pandemic, on the canals that cut through the Venetian island of Burano, where the photographs of them that went viral were taken. The images of the dolphins, meanwhile, had been snapped not in Venice but at a port in Sardinia. The elephant item, similarly, was discredited.

That these specific news items were fake is not, in a sense, that important. For one thing, it would be a logical fallacy to jump from disproving their veracity to concluding that other-than-human animals were not indeed thriving in a world from which humans had taken a major backward step. One only needs to spend a few minutes outdoors, even in an urban setting, to confirm that many animals are wary of humans, and so it stands to reason that an absence of humans would be exploited in some way – that the humanless vacuum, in other words, would be filled. For another thing, the fact that these news items went viral showed that people, in troubling times, were receptive to stories not just of hope in general but of hope for our non-human kin. This was just one demonstration of what, in the authors' experience, was a much larger phenomenon: the lockdowns of 2020 piqued a societal interest in the more-than-human world.

During the early stages of the Covid-19 pandemic, acquaintances (among our respective groups of friends and colleagues) who had not previously shown more than a passing interest in the wild nature around them were sharing observations of non-humans in a way that they never previously had. And those who already had a moderate interest were showing signs of developing a deep enthusiasm. We can speculate on some of the manifold proximate causes of such occurrences. First, with a greatly restricted set of activities from which they could choose to engage in, many people found their lives to be suddenly simplified, and in the absence of competing 'life distractions' had time to notice nature. Secondly, with the skies mostly clear of planes and the roads carrying far lighter traffic than they normally would be, people could hear the sounds of their wild neighbours with an unprecedented clarity. Indeed, one of the most frequent comments on urban wildlife that we encountered during this time concerned the wonder of being able to experience birdsong in its full majesty; our avian kin were, for once, not being drowned out by the incessant noise of machines. Thirdly, many people were exploring their local areas on foot or bicycle like they never had before, and – being grateful simply to be outside and breathing fresh air – were especially receptive to goings-on around them, including those that were other-than-human. Fourthly, with social-distancing in place and unprecedented levels of digital-only contact between humans, people, we conjecture, were craving real, in-the-skin encounters with other beings, and they sought these, in the absence of human company, from wild non-human animals and plants.

Most importantly, in regard our discussion here, all of the signs described above suggest, to us at least, that there is a real possibility of a society-wide revival of natural history as an interest and passion. In some ways, this is hardly surprising, as the practice of natural history is in our blood. (To bowdlerise Harry, the London gangster played by Ralph Fiennes in the film *In Bruges:* "How can fricking swans not fricking be somebody's fricking thing?") It has served humans well throughout the history of our species, guiding us in our foraging and hunting, our danger avoidance, our medicine, our shelter-making, and our ceremonies. And it is only in the most recent generations, when a proportion of the population has been able to lead successful lives in ignorance of plant-gathering and agriculture, that it has been possible to *get by* without an interest in natural history.

But *get by* they have done. Over the past generation or two, there has been a strong tendency for the lives and daily pursuits of humans to become decoupled from goings-on in the rest of nature. This has occurred in children's outdoor recreation, as the norms of society in regard to play have shifted (not helping here is the loss, in the past few decades, of natural history education from most school curricula, and the replacement of nature-oriented summer activity programmes with sports camps, computer camps, and the like). The decoupling has also happened in the technologically cosseted lives of many Western adults. Here, as was observed in 2011 by Tom Fleischner, founding President of the Natural History Network, in growing older "we have to learn to *not* pay attention to our world" (Fleischner, 2011: 21). Advertising and the consumerist culture, as Fleischner has commented, are major forces in shrinking the scope of our attention.

The decoupling is even apparent in the teaching and practice of biology, from compulsory education through to institutes of higher education. "Despite the importance of detailed natural history information to many sectors of society," as Tewksbury and colleagues (2014: 304) have remarked, "exposure and training in traditional forms of natural history have not kept pace with growth in the natural sciences over the past 50 years."

The journalist Richard Louv is someone who has dedicated much of his life to researching and writing on this issue of decoupling. In *Last Child in the Woods* (Louv, 2008), he shined a spotlight on the divide that has grown in modern times between children and the outdoors and drew links from the resulting 'nature-deficit' to increases in such phenomena as obesity, depression, and attention disorders. Then, in *The Nature Principle* (Louv, 2012), he shifted his focus to nature-deficit in adult lives and argued how the restorative powers of the natural world can enhance health, promote creativity and mental acuity, foster better businesses and communities, and strengthen the bonds between humans. The deficit of which Louv has written is, of course, not a material one – humans are just as dependent as they ever have been on the workings of nature around them, for breathable air, freshwater, food, and so forth – but, rather, a deficit of spirit and attention. The importance of each of the relationships that he has described between contact with non-human nature and human wellbeing cannot be overstated. Yet, from an ecocentric (Earth-centred) perspective, there is something that makes the disconnect more troubling still, as we explore below.

Natural history as a wellspring of care for non-humans

Before getting any deeper into this chapter, we should come to defining 'natural history'. The term has been used in various ways through the centuries, and we will pick out two examples here, both contemporary. The first is that of Joshua Tewksbury and colleagues (2014: 300), who have described 'natural history' as "the observation and description of the natural world, with the study of organisms and their linkages to the environment being central." This is both crisp and sound as a definition. But the second representation that we select – drawing on the work of Tom Fleischner once more – is the one that we prefer, for its greater emotive potential. In a 2002 article in *Wild Earth,* Fleischner defined natural history as the "practice of intentional, focused attentiveness and receptivity to the more-than-human world, guided by honesty and accuracy" (Fleischner, 2002: 11).

Building on this, Fleischner has commented elsewhere that "natural history facilitates people falling in love with the world" (see https://is.gd/nathist). And "a known and loved world," he reflected in that *Wild Earth* piece, "has more effective advocates than one that's ignored" (Fleischner, 2002: 12). The mechanism for this is simple: people spend a part of their lives observing and learning about their fellow creatures; they develop affection for, and feelings of kinship with, these beings; and they, in turn, want to see that the creatures and their habitats are protected from harm. This begins with striving to behave, as individuals, in ways that do not compromise the flourishing of the wild species that one has grown to cherish. It also requires continued immersion in nature to renew relationships with those wild beings and to renew one's energy for the long fight ahead (see Figure 1).

It is for the reasons presented in the previous paragraph that the disconnect between humans and non-human nature, as we noted above, is particularly troubling from an ecocentric perspective. In Richard Louv's most recent book, *Our Wild Calling* (Louv, 2019), he reaches beyond the inwardly human focus of his previous books and looks to the importance of repairing the disconnect not just for our own species but for all life. He writes, for instance: "The only way people come to truly care about animals is to know them, to immerse themselves in the flow of nature and the lives of animals" (Louv, 2019: 265). He also suggests the following 'reciprocity principle': "For every moment of healing that humans receive from another creature, humans [should] provide an equal moment of healing for that animal and its kin" (Louv, 2019: 272"). There is much wisdom embedded in this proposal. Additionally, the study of natural history contributes directly to conservation because, in order to save species, we must know something about their habitat requirements, their life histories, and their habits, which can only be gained through direct observations in the wild.

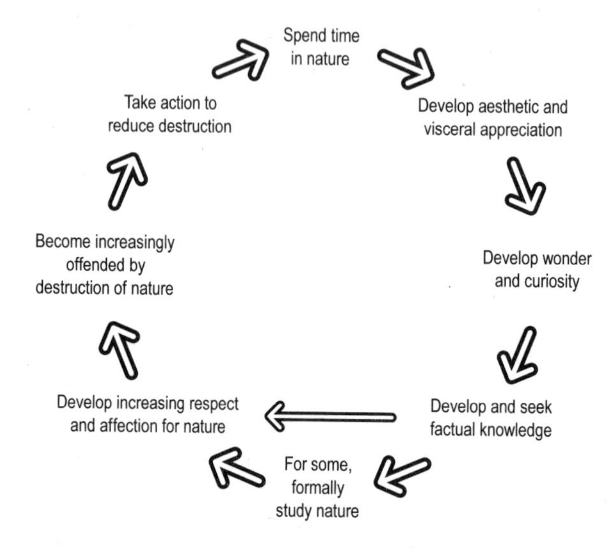

Figure 1: Spending time in nature and studying natural history lead to ever-increasing appreciation, wonder, knowledge, and respect for nature, and greater dedication to conservation. Continuing the loop, spending time in nature provides emotional breaks from the 'gloomy business' of conservation and renews acquaintances with wild beings. From *Forgotten Grasslands of the South*, by Reed F Noss (copyright © 2013 Island Press; reproduced with permission of the publisher).

In short, then, the potential benefits of the *practice of intentional, focused attentiveness and receptivity to the more-than-human world* go beyond helping with health, creativity, and societal functioning in humans to offering, more crucially still, a wellspring of informed care for our non-human kin and the habitats that they depend on. This is exactly the kind of care that is so desperately needed if humanity is to reverse its current course of 'ecocide everywhere'. It is no exaggeration to say that the study of natural history can help us save the Earth.

What might drive a revival?

"Never in history," muse University of Vermont lecturers Matthew Kolan and Walter Poleman (2009: 30), "have we stood where we do now – with the knowledge, technology, and power to fundamentally alter the geological, biological, cultural, and atmospheric processes upon which we rely for survival." And at "this critical moment in time," they reflect, "we are in need of a different approach to education and learning – one that reveals connections, strengthens relationships, and recognises the whole." A core part of such an educational approach, as Kolan and Poleman themselves note, must surely be a strong focus on natural history. This is as true for teaching that is conducted in a formal setting as it is for education within the context of life-long learning. In other words, a society-wide revival of natural history lies at the heart of any deep solution to the Earth's, and humanity's, currently dire predicament. For natural history has the potential to be so much more than a mere interest, or a stimulating distraction from goings-on in everyday life. Rather, its focus *is* everyday life. And, if taught right, it is inherently fascinating to people because it stimulates their innate biophilia – their emotional connection to other living beings.

Natural history, when understood as *intentional, focused attentiveness and receptivity to the more-than-human world,* is a practice that transforms the abstract into the visceral. Natural history also turns objects into subjects, subjects that – to quote field naturalist Ian Whyte, in an article that he co-wrote with one of us – are "imbued with meaning and value and that have independent concerns" (Whyte and Gray, 2020: 119). Natural history, furthermore, reveals the interconnectedness of all life and, as such, has a large role to play in repairing humanity's severed connection with the rest of nature. Natural history, relatedly, shows us our true place in nature: one of countless co-functioning elements, not master of the machine. Finally, in regard to the desolation that humanity is currently leaving in its wake, natural history has the potential to confine ignorance to the cellar of expired excuses. No one who knows but does not act is innocent.

So, what could, and should, natural history's revival look like? Firstly, and perhaps most obviously, it is crucial that people have a chance to connect in deep ways with the nature around them during the formative years of childhood. There are already many superb organisations around the world, from the local level to the international, working to provide children with these chances (including, to cite just one, Richard Louv's own Children and Nature Network). To make access to such opportunities close to universal, though, it is essential that natural history be incorporated within school curricula. And, by incorporated, we mean not as a side activity to offer relief from the main programme – as is typically the case at present, where, occasionally, it is taught – but as a fundamental aspect of the educational agenda, given as much attention as literacy and even more, dare we say it, than numeracy. Natural history as a subject need not be structured around the teaching of examinable knowledge, something that has the ultimate purpose of incentivising educational participation by the ultimate ranking of participants. Instead, the opportunity to leave the desk behind and enter an outdoor classroom is likely, for most children, to be a sufficient incentive, in itself, for meaningful participation. Most children will also delight in the opportunity to put their textbooks to one side and engage their full array of senses – watching birds collect twigs for their nests, smelling crushed conifer needles, touching the smooth bark of a beech tree, experiencing the pleasantly peppery taste of hedge garlic, and listening to the unseen crickets chirr. Finally, in making observations, keeping field diaries, and discussing their thoughts and findings with others, these children will have a chance both to practise and improve literacy and numeracy skills in a novel, non-examined context, and to develop important social skills in an informal setting.

The above discussion is not intended to preclude formal examination-based courses for natural history. And there are already positive moves afoot in this area. In the UK, for instance, the OCR exam board announced in 2020 that it was consulting on the launch of a GCSE in the subject (GCSEs are key exams taken at, or close to, the end of compulsory schooling; OCR, 2020). This course is something that the writer Mary Colwell has been campaigning for since 2011.

There is also much scope for enhancing the place of natural history within the biology that is taught and practised in institutes of higher education, the decline of which was noted above. Nevertheless, we offer a cautionary note that the attainment of grades and qualifications has the potential to run counter to the spirit of natural history. The best practitioners of the subject – and, happily, they form the majority in our experience – are the ones who realise that natural history is not about them, but, instead, the countless other lives that make up our Earthly cosmos. These are the natural historians who return from trips to new locales not boasting of species ticked off a list, as proof of their own ability, but with their spirit recharged, their awareness-of-others expanded, and their humility enhanced. These are the natural historians who are deeply grateful to have been able to enter the habitat of such graceful fellow beings. Furthermore, the need for a society-wide revival calls for natural history to become a fundamental aspect of the educational agenda, and not just a narrow avenue in which certain students get one more qualification, or one more grade, to add to their list. In short, then, we commend ongoing efforts to develop formal qualifications in natural history, and we urge those seeking to restore and expand the place of natural history within university programmes to continue their important work, but we note that these developments *by themselves* will not foster society-wide change.

Now we come to the second part of the grand revival of natural history. This will be, very simply, a suite of programmes for adults, who form a vast swathe of the population that is often overlooked in the targeting of nature activities (other than as accompaniers of children in family-oriented projects). Again, there is of course work that is already being done in this area. For instance, the conservation projects of today, especially those in urban areas and other populous regions, often comprise elements of adult-focused public outreach and volunteer engagement. There is also an increasing number of citizen-science initiatives, which enlist the help of adults in recording sightings of birds, butterflies, and other wildlife. As valuable as such activities are, though, they do have a tendency to attract people with an established interest in natural history. In other words, they mostly reach only the 'low-hanging fruit' or – to add wear to another overworked phrase – preach only to the converted.

One way in which the tendrils of adult-facing programmes can span across a broader cross-section of society is for outreach sessions and materials to be pitched at an introductory level and presented to 'captive audiences' whose existence is not related to natural history. Such groups include those of companies who offer their employees team-building events and other such training activities, those of civic organisations such as the University of the Third Age, and those brought together by churches, mosques, temples, and other places of worship. Another way – one that would potentially be more effective still – is for adult programmes to be scaled up through delivery in partnership with, and drawing funding from, local government. There are many good reasons why the political administration of a city, say, would wish to make natural history a high priority. One such reason is the mental health benefits that can result from increased contact with nature. Improving the mental health of individuals would be a good thing in its own right but could also reduce the burden placed by modern society on already-creaking health systems. Another reason, we believe, is that natural history education could help inculcate green behaviours among a citizenship, which are behaviours that ultimately reduce the stress placed on water supplies, refuse-disposal operations, and other key municipal services.

Finally, the power of advertising – which itself, as noted above, can be a major distraction from the world of beauty around us – could, in this context, be harnessed for good. We are not aware of any campaigns for fostering an interest in natural history that have been run specifically to attract the attention of people with a baseline interest of close to zero. But given how effective advertising can be at selling things that people really do not need, there is no reason to suppose that it would not work for something – contact with nature – from which people would draw genuine and important benefits. To give an example of how this might work, a conservation charity could run a non-targeted campaign of adverts across print and online newspapers that was linked to simple supporting materials for suggested introductory activities, such as putting names to local butterflies.

A note on recreational pressure

We cannot speak of a grand revival of natural history without asking what the negative impacts of increased visitor numbers might be on wild places. Many wild beauty spots are already placed under tremendous pressure by the presence of human visitors. Erosion, trampling, excessive noise, littering and other types of pollution all contribute to making such wild places less habitable by the creatures who need them. "Are we loving our national parks to death?" was the question fittingly posed by Dayton Duncan in the title of a 2016 piece in *The New York Times* (Duncan, 2016).

There are three comments that we wish to make about this issue. The first is that it is reasonable to assume that those practising natural history – as *intentional, focused attentiveness and receptivity to the more-than-human world* – will most likely want to avoid doing harm to the places they visit. The same cannot necessarily be said of people visiting famed scenic outlooks with the over-riding purpose of getting a photo of the place (or themselves with the places as a backdrop) simply to publish, often boastfully, on social media. The rise in this latter phenomenon and the associated problems from the increased recreational pressure that it brings to bear have been examined by Charlotte Simmonds and colleagues (2018). With a nod to the title of Duncan's above-mentioned piece, they headlined their own analysis: "Crisis in our national parks: how tourists are loving nature to death."

The second comment is that, as postdoctoral researcher Desiree Narango (2020: 13) has noted, the "easiest and most accessible place for people to connect with nature is where they live." And since "the majority of people now live in urban and suburban areas," she writes, such places, and not wild beauty spots, are where most "primary interactions with nature and wildlife are occurring" (Narango, 2020: 13). (Narango has herself described an example of a project combining natural history, urban ecology, and citizen science, using stories that, in her words, "can encourage people to care enough to do something about conserving biodiversity at home" [Narango, 2020: 16].)

A third comment regards the need for increased land conservation. In order to facilitate human interactions with nature without overwhelming the wild, it is necessary for land conservation programmes to be greatly expanded. This ranges from protection of undeveloped areas within city limits, outward to the wildest remaining landscapes (for more on this, see Chapter 3, by George Wuerthner). With more protected natural and semi-natural area, the impacts of human visitation will be diluted.

The role of technology: From cause to solution?

The technological advances of the past century are the principal cause of the mass decoupling of humans from the more-than-human world. Yet it would be folly *not* to explore what role technology might have in helping to at least partially fix the predicament of detachment that it is has driven. Several major positives for the practice of natural history have arisen, for instance, from the creation of the internet. One is the wide availability of free-to-access resources for helping to identify species encountered and to learn something of their ecology, removing the potential barrier of cost that comes with books (printed materials are still very valuable in this area, but at least there is now an alternative of sorts). Another major positive is the ease with which individuals can now submit records of the species that they have encountered for verification and inclusion in databases that help inform conservation in practice. This activity not only supports the learning process but also gives a greater purpose to making notes of what one sees. In parallel, the progress towards ubiquity of the smartphone, and the improvements in the quality of their integral cameras, means that many people now have on their person, by default, a means of capturing images of plants, insects, and other easy-to-photograph organisms. These images can be useful both in aiding one's self-education and in supporting records, some of considerable scientific value, that are submitted to databases. (In the UK, for instance, records can be submitted through the iRecord initiative at https://www.brc.ac.uk/irecord/.)

Modern technology also offers the potential for virtual 'contact' with nature. This can be as simple as watching nature documentaries on television. While these programmes do have the potential to be misleading as to the current state of the planet, offering a false reassurance that all is well, they also have great potential for education in a medium that appeals to many people. And it is hard to argue against the idea that it is better, all told, to have a small film crew visit Antarctica, say, than to have millions of people choosing to see penguins in person.

More immersive experiences are available in the form of nature-based computer games – including *Shelter* (a 2013 creation by Might and Delight), in which the player experiences the wild as a female badger protecting her cubs from a suite of hazards – and in digital experiences such as the Ocean Odyssey. The latter is a National Geographic Encounter digital aquarium in which visitors are able to interact, virtually, with creatures of the sea. Such interactive media have a genuine potential, in the words of Alf Seegert (2014), a professor of English at the University of Utah, to "evoke empathic identification" with non-human others. And such empathy is at the heart of what one might call platinum-level natural history.

We would be the first people to note that virtual 'contact' is, in an important sense, not comparable to experiencing real nature, and that it can only ever be a supplement rather than a true substitute. At the same time, we are not going to prescribe a single format by which everyone has to develop an interest in natural history, and, at a moment in time in which mass extinction fills the horizon, we do not wish to take anything with potential for good off the table.

Concluding remark

What we are calling for in this chapter represents nothing less than a seismic shift in modern society. As such, we are open to accusations of being dreamers who lack a grounding in reality. Our goal here, though, has been to present a vision, and the best visions are untempered and undiluted. We make no apologies for having offered such a vista, and we are quite certain that it represents true reality.

References

Daly, N. (2020). Fake animal news abounds on social media as coronavirus upends life. *National Geographic* (22 March). Available at https://www.nationalgeographic.co.uk/animals/2020/03/fake-animal-news-abounds-social-media-coronavirus-upends-life (Accessed September 2020).

Duncan, D. (2016). Are we loving our national parks to death? *The New York Times* (6 August). Available at https://www.nytimes.com/2016/08/07/opinion/sunday/are-we-loving-our-national-parks-to-death.html (Accessed September 2020).

Fleischner, T. L. (2002). Natural history and the spiral of offering. *Wild Earth 11,* p. 10-13.

Fleischner, T. L. (2011). Why natural history matters. *Journal of Natural History Education 5,* p. 21-24.

Kolan, M. & Poleman, W. (2009). Revitalizing natural history education by design. *Journal of Natural History Education 3,* p. 30-40.

Louv, R. (2008). *Last Child in the Woods: Saving our children from nature-deficit disorder.* Chapel Hill, NC: Algonquin Books.

Louv, R. (2012). *The Nature Principle: Reconnecting with life in a virtual age.* Chapel Hill, NC: Algonquin Books.

Louv, R. (2019). *Our Wild Calling: How connecting with animals can transform our lives—and save theirs.* Chapel Hill, NC: Algonquin Books.

Narango, D. L. (2020). Natural history in the city: Connecting people to the ecology of their plant and animal neighbors. *Journal of Natural History Education 14,* p. 13-17.

OCR (2020). *Re-connecting young people and nature.* OCR News (5 June). Available at https://www.ocr.org.uk/news/re-connecting-young-people-and-nature/ (Accessed September 2020).

Seegert, A. (2014). Pixels and pathos: Video games and empathy. Presented at: *Interdisciplinary Symposium on Empathy, Contemplative Practice and Pedagogy, the Humanities, and the Sciences,* University of Utah, UT.

Simmonds, C., McGivney, A., Reilly, P., et al. (2018). Crisis in our national parks: how tourists are loving nature to death. *The Guardian* (20 November). Available at https://www.theguardian.com/

environment/2018/nov/20/national-parks-america-overcrowding-crisis-tourism-visitation-solutions (Accessed September 2020).

Tewksbury, J. J., Anderson, J. G., Bakker, J. D., et al. (2014). Natural history's place in science and society. *BioScience 64,* p. 300-310.

Trombulak, S. C. & Fleischner, T. L. (2007). Natural history renaissance. *Journal of Natural History Education 1,* p. 1-4.

Whyte, I. & Gray, J. (2020). Field guides as a gateway to appreciating more-than-human concerns. *The Ecological Citizen 3,* p. 119.

Every one of us can make a contribution. And quite often we are looking for the big things and forget that, wherever we are, we can make a contribution. Sometimes I tell myself, I may only be planting a tree here, but just imagine what's happening if there are billions of people out there doing something. Just imagine the power of what we can do.

Wangari Maathai
Founder of the Green Belt Movement

Chapter 6:
Case Study 2: Indigenous Teachings in an Early-Childhood Programme on the Banks of the GabeKanang Ziibi

Editor's note: In this brief case study, we present extracts from an article that was published during 2020 about an inspiring early-childhood programme being taught in forests and meadows on the banks of the GabeKanang Ziibi (the Humber River), outside the Canadian city of Toronto. The programme promotes Indigenous wisdom, starting with the Anishinaabe teaching that Nibi (water) is the blood of Aki (the Earth).

The article from which these extracts are drawn was: Zimanyi L, Keeshig H, and Short L (2020) Children make connections to Aki (Earth) through Anishinaabe teachings. *The Conversation*, 19 April. Available at: https://theconversation.com/children-make-connections-to-aki-earth-through-anishinaabe-teachings-133669 (Accessed February 2021).

First, noting that 'water tables', which provide opportunities for sensory water play and learning in children, are a part of many mainstream early-childhood programmes, the authors of the article describe how their programme extends beyond this and takes education outside.

"In The Willows, a land-based early childhood program at

Humber College in the northwest end of Toronto, engaging with water is extended outdoors — in forests and meadows, by the wetlands, ponds and GabeKanang Ziibi (the Humber River).

"On the traditional and treaty lands of the Mississaugas of the Credit, young children interact with water through the seasons and in all weather. Playing in the rain offers new experiences and perspectives; water is also a gift for the trees, birds and the river."

As part of this, Lynn Short, one of the authors of the article, regularly walks with children on the land and shares Anishinaabe teachings and stories from Onaubinosay Elder Jim Dumont, who "teaches that water is medicine."

Children, the authors note, are naturally drawn to these outdoor activities:

"They throw stones back to the water, jump in puddles.

"They notice turtles on logs, investigate velvety green mosses revived by spring rains and make up stories about muddy animal tracks."

Land as teacher

A key aspect of the early-childhood programme is seeing land as teacher, as the authors explain:

"In seeing land as a teacher that supports respectful, relational, reciprocal and responsible engagements, children learn about relationships with Neekaanagana (All Our Relations), like land, water, animals, plant life and other beings that human beings depend upon for survival.

"We learn the ways of the ones who take care of us, so we may take care of them.

"With Indigenous knowledge holders and storytellers, we walk together and learn that each element, plant and animal not only has a name but a spirit, a story, a gift and a responsibility, as do we humans."

Reverencing water

One of the aims of the programme is to build community relationships with people from Indigenous cultural networks. Helena (Joanne) Keeshig Joanne, one of the authors of the original article and person with whom the programme has connected, says this about the importance of reverencing water:

"Being respectful of and having respect for water is important. Experiencing and connecting to water in its natural environment is essential for children to develop confidence, master skill[s] and overcome fear. Building a healthy relationship not only with the environment, but with all those who call the land home is crucial as those resources sustain us."

Restoring relationships

A final facet to this programme is the restoring of relationships. As is explained in the article:

"In acknowledging nature as teacher, in our early childhood program, we ask ourselves what we can do to honour our relationships that allow all of Earth's creations to live in harmony and balance.

"We read Nibi Emosaawdang (The Water Walker) about Anishinaabe Elder and Nokomis (grandmother) Josephine Ba-Mandamin who walked to protect the waters of the Earth.

"We share and draw stories about the water."

That goal of honouring relationships that allow all beings to live in harmony should be at the heart of every nature education programme.

As you walk look around, assess where you are, reflect on where you have been, and dream of where you are going. Every moment of the present contains the seeds of opportunity for change. Your life is an adventure. Live it fully.

John Francis
Author of *Planetwalker*

Chapter 7:

Back to the Future: Ecovillages and Eco-Communities – a Solution to Work, Technology and Climate Change Challenges

Andrew Olivier

Covid-19 has paused economies, temporarily stopping the growth mantra. Our species progress to date has been devastating to the planet and its biodiversity. Climate change poses a real and present threat. World leaders appear to flounder in the complexity of work required, while young girls have louder, more urgent voices. Sadly, COP meetings are more about creating new carbon markets than ridding us of carbon. Few of the decision makers at COP 25 listened to the nature-based solutions, the theme of that gathering.

The UN Sustainable Development Goals continue to limp along, testimony to the fact that more and more people are aware of the need for real change, but lacking legislation to enable the change. No world leader wants to risk damaging the growth economy. Climate change is insidious and escalating, but Covid-19 stepped in and stole the show, a sort of precursor to the future, putting the spotlight on how fragile we are.

Fortunately, there are good solutions and some are already being implemented. Let's look at one of them, starting with the basics.

The importance of work

Maslow's hierarchy of needs indicates that at its most basic, work is a means to provide security. For a significant part of the world's population, it still is. While standards of living have lifted, the world is still a tough place. Almost a billion people continue to live below or just above the poverty line and for a billion more, security is tenuous. We must not think that we have moved that far from meeting our basic security needs. Even for the well-off, we are still three pay checks away from disaster. Covid-19 has illuminated this, by pushing poverty levels back 30 years, impacting 90 million people (1).

For those who have managed to push through the security issues, the lure of esteem and self actualisation allows us to focus on that intriguing notion of our personal purpose. We like to feel valued, that we are contributing; doing something useful; that is why we work, paid or not. We actively seek meaningful work linked to our personal purpose. It's the stuff we like to do... that we are good at and which gives us satisfaction.

Work is where we encounter that magic band of energy, called "Flow" (Csikszentmihalyi, 1990). Flow is where we feel connected to our work, where time disappears, where our skills, education, knowledge, values and cognitive abilities are engaged. Decision making is fulfilling, we love the challenge, we feel able to make a valuable contribution and are recognised for our efforts. Just think back to your own working journey and the times you really loved. Flow is linked to the actual work we do, which can be grouped into generic themes of work. Our flow band shifts as we mature, so for some, the work we do in our twenties, is not rewarding in our forties. Sadly, many people have never experienced flow. They do drudge, back breaking work because the conditions of security is not met. They survive, not thrive.

Waves of change

The First Wave, the Age of Agriculture, heralded the growth of towns, cities, trade, population, diversification of work, the rise and fall of empires and the growth of major organised religions. The pace was slow, technology and communication basic. Life was lived according to the season and the setting of the sun.

The Second Wave, the Industrial Age, was accompanied by bloody warfare as First and Second Waves clashed (American Civil War, Anglo-Boer War, Russian Revolution, Mao's Cultural Revolution) for minerals, markets and manpower. The age of 'mass' arrived: mass production, mass consumption, mass communication, mass education and it required masses of educated workers and consumers. Market capitalism saw the dollar replacing religion. The growth of the state heralded national centralisation and consolidation. Horrific wars of mass destruction (World War I and II), culminating in the ultimate weapons of mass destruction.

Nature was a force to beat or bend to our will. Improved health care and science saw the world's population and city sizes burgeon. Complexity of business, government and governance grew. People stopped being producers of food and became consumers. Generations raised in the urban milieu, lost their roots. Nuclear families replaced communities. Colonialism, debt and wage-earning replaced slavery. These waves of change washed unevenly across the world. Power, wealth and control belonged to those at the forefront.

Yet waves of change continued, intensifying. The Third Wave heralded the dawn of Information and Technology. Power has subtly shifted to global data behemoths, and we are witnessing revolutions in information and biotechnologies. Many do not fully understand the ramifications of this new Age. What is clear is that the new technologies of AI, automation, machine learning and the mighty algorithm is anticipated to wreak havoc on jobs and careers. We are warned to expect huge job losses in the unskilled, skilled, professional, and managerial work ranks. Fourth Wave has arrived.

Jobless job markets

Some estimate between 35-50% of current jobs may disappear by 2040. In the past job/occupational losses have been replaced by new jobs and new occupations. Not so now, Harari warns, this is different. Business is shedding jobs in the digital age. New roles and occupations are open to those with access to elite education. And for them, the future is one of continually upskilling and diversification as technology marches ahead. For the youth, in many 'developed nations' land ownership is out of reach and the diminished gig economy awaits. At the same time governments demand new jobs be created.

The Fourth Wave may create the most polarised society we have ever seen. The world of work is being reconfigured for the elite. In many ways the new technologies support this. Harari suggests we may end up with a surplus of irrelevant humanity. Unemployable, unable to meet Maslow's hierarchy, even at the most basic level, living in failing states where globalisation has retreated, and jobs have disappeared.

Covid-19 arrived as the Fourth Wave got into swing. Many in the developed world have been lucky to receive government support, but many more do not have that privilege. The issue of universal basic income reared its head briefly, but governments balked at the cost of debt. The pandemic has shown no one is safe from job losses and how vulnerable cities are to disruption.

The science of doing stuff

More than half a century of research into how people work has generated some interesting principles, with strong links into complexity science and complex adaptive systems (Craddock, 2014). A basic principle is that work tends to increase in complexity (the number of variables, the rate of change, the difficulty of identifying the parts) and uncertainty as it gets more complex. Table 1 below indicates this. People find flow in specific themes of work complexity as they mature. Our ability to handle complexity changes as we mature. The key is where do one find satisfactions or flow.

Table 1: General categories and principles of Work[1]

Work Theme	Example of Work Types	General Principle of Work
Quality	Findhorn Ecovillage co-worker & Team Leader, waitron, Uber delivery, front line sales person, call centre operator, police constable, artisan, technician, CAD Engineer, claims assessor, miner, chef, front line supervisor, Agile Team Member (Squad), estate agent	Direct work – produce tangible results in a given time. Ranges from unskilled to fully skilled work. *Time Span of up to three months.*
Service	Professionals/Specialists (e.g. doctor, scientist, accountant), front line manager, teacher, area sales manager, head chef, systems analyst, permaculturalist, Agile Team member (Squad), supermarket outlet, small business owner, Findhorn work team focaliser.	Research, analysis of specific issues; manage team doing direct work to achieve specific outcomes, generate solutions or options. *Time span of one year.*
Practice	Manager, project leader, research programme leader, human resource manager, academic department head, practice head, logistics manager, colonel, ship's captain, call centre manager. Findhorn Ecovillage Line Manager or Area Focalisers, Department Head, Network Coordinator, Steering Circle Lead (sociocracy) in large ecovillages (Narara Ecovillage).	Coordinating, integrating, blending components of a system to meet goals, objectives. Constructing, connecting fine tuning systems to make the most of resources. Tactical strategies. *Time Span of two years.*
Strategic Development	Dean, general manager, principal advisor, senior consultant, one-star general, deputy director, project managers on complex projects, CEO. C-suite of complex business. CEO/MD/Leader of large ecovillages	Coordinate and resource multiple programmes. Modelling new futures, services, technologies, products, position and transition old to new. *Time Span of up to five years.*
Strategic Intent	CEO of a large, complex company, MD of division, director general, vice-chancellor of university, two stars general. Governing Circle of Large Ecovillages, Trustees /Community	Direct/coordinate the direction and intent of a unified system – setting direction, sustainability and viability of a complex work system. *Time Span of up to ten years.*

An estimated 95% of the world's population find work in the themes of Quality, Service and Practice. Statistically, our data would indicate that this is where most people find flow. It is precisely in these Work Themes that we are most at risk.

To tweet or not to tweet (short termism)

Markets reward short term profitably. The idea of thinking and planning for twenty or fifty years hence is labelled old fashioned and unrealistic. It is justified by saying technology is changing too fast. Research has identified a link between Time and Complexity

1. This table is compiled from various sources. It is not complete, there are two other work themes not indicated with time spans of up to a hundred years. They are not well represented at the moment.

in decision making. Notice Time Span, right hand column of Table 1 – this refers to the amount of time that lapses before we see the outcome of our most complex decisions. Business has ridiculed this idea, but we are now becoming aware of the horrors of failing to think long term (2) and ignoring multiple capitals in our costing. Instant gratification and connectivity detracts from our ability to focus on more complex issues, so we are becoming dumber (Newport, 2016).

Oh, Climate Change?

So once we have the vaccine (or as the WHO says, we become Covid-19 ready societies) will we rush back to where we left off with some minor changes? A Fifth Wave is needed, that of living in an ecologically sound way, in harmony with nature and living within the carrying capacity of the Earth. Sadly climate change is not seen as a threat, generating a reaction like Covid-19.

Living differently: a mindshift followed by behaviours

Global awareness of our predicament is increasing. Support is growing through activists, campaigns, and networks to address issues global leadership is unable to do. Networks are springing up to address diverse causes, from the macro to the local. It is about people taking back power, building sustainability, resilience, and a slower, happier world, and for many, it is not wanting to play the market capital game.

An alternative future is here already

Eco-communities have already made the transition to a new Fifth Wave of sustainable living. Growing numbers are interested in this trend. An example of this is the Global Ecovillage Network or GEN for short, which has grown organically over the last 25 years, into 5 regional networks of ecovillage and eco-communities (South America, North America, Europe, Africa, Asia and Oceania and the youth arm, NextGEN).

An ecovillage is an intentional, traditional or urban community that is consciously designing its pathway through locally owned, participatory processes, and aiming to address the current crises through the Ecovillage Principles in the 4 Areas of Regeneration (social, culture, ecology, economy into a whole systems design) (3).

Ecovillages are living laboratories pioneering alternative and innovative solutions. They are often vibrant social structures, vastly diverse, yet united in their actions towards low-impact, high-quality lifestyles. Ecovillages are complex settlements with local economies offering a range of different services to suit community needs. For example;

Health and care for the elderly

Covid-19 has illustrated the weaknesses of current aged care facilities in many countries. Some ecovillages have pioneered a whole systems approach to aged care. My first exposure was at Aldinga Ecovillage in Australia, where I was invited to the bedside of a terminally ill patient. The community had rallied around, offering the opportunity to transition at home, with meals and respite afforded his partner, and being surrounded by love and support.

On a larger scale, the Heilhaus community in Kassel (Germany), works with the circle of life holistically. Its slogan is Birth, Life and Death. The founder, a spiritual healer, offers weekly public healing sessions. Heilhaus has grown into a community of over 350 people, living in eco-friendly flats, with dedicated business buildings and a striking community centre. Their economy is built around a midwifery, a school for severely challenged children, a multi-generational palliative care center and a day clinic, with regular medical practitioners, carers and alternative therapists. A canteen provides organic meals for workers, guests and visitors. The community is employed in the business and the centre has a wide spread reputation as a place where alternative health and modern science meet.

Rural regeneration

There are still hundreds of thousands of traditional indigenous villages in the world. These communities, themselves already ecovillages, are under huge pressure due to a myriad of factors. For example; urbanisation, poverty, land grab, corruption, conflict, environmental degradation, social and cultural pressures. Community leaders are increasingly vocal in trying to shift into a regenerative cycle to rebuild their communities. So it is natural for other ecovillages to be deeply focused on regeneration.

REDES on the border of Senegal and Mali is part of The Great Green Wall, an African-led movement with the inspiring ambition to grow an 8,000 km natural wonder of the world across the entire width of Africa. The initiative is already bringing life back to Africa's degraded landscapes at an unprecedented scale, providing food security, jobs and a reason to stay for the millions who live along its path. REDES is working for the regeneration of the socio-cultural and physical environments of the Sahel to generate wealth for local communities and establish peace and harmony in the region.

This project is currently transforming 41 rural villages, close to the River Senegal. It includes 50 Senegalese villages and 50 Mauritanian rural communities. REDES programmes include running green caravans, implementing community orchards, organising food processing workshops, digging wells for drinking water, building classrooms in villages, cleaning up villages, organising free medical consultations, film projections, conferences, facilitating academic research and offering practical ecovillage design education.

Another network member is Sarvodaya, an NGO, working with some 2,000 active sustainable villages in Sri Lanka. In India, INSEAD works with a range of communities in the transitioning to ecovillages along the four principles of sustainable education, supported by appropriate technology, while Auroville is an eco-city of some 3,500 people that has its own internal economy.

Education

Other ecovillages have built their unique value add around teaching, education, shifting mindsets and behaviours. Offerings may range from practical skills of sustainable farming, green building, how to live together harmoniously, conflict management, conference hosting, collaborative decision making to spiritual awakening. The spectrum is broad.

The strength of GEN lies in its education programmes, of bringing strangers to these distributed networks of learning and exposing them to life in communities. Research would indicate this works. A recent study has shown how in Brazil, exposing students from different disciplines to ecovillages living profoundly shifts their view on how to be sustainable and that it made them think differently (4). Covid-19 has seen GEN innovate with online offerings and webinars, the young embracing the new technologies. No luddites here.

Run with wolves

Nashira in Colombia, is an ecovillage for women (5). It was created to balance the results of a long civil war and its societal impact. 75% of Nashira's households are headed by women as they are the main income earners. All administrative decisions are taken by women through consensus. Men who have initiated violent acts against their partners or children have been expelled. There is no crime. Violence against women was eradicated. It is a community with open doors, where women support each other, and men have developed a new culture of love and respect for them. Childcare and maintaining the ecovillage are tasks shared through collective work.

The women of Nashira cultivate staple crops using permaculture techniques, fruit trees flourish in the common areas used by the whole community. Some of the cooking uses solar power and the women proudly collect and recycle organic and inorganic waste from the neighbourhood. The recycled plastic, glass and other inorganic materials are used to make products for their own use. They welcome visitors and host conferences. A model for other parts of the world?

The glue that binds...

Ecovillages represent beacons of hope for living differently. Human relationships are a common denominator, with song, dance, music and ritual binding individuals and connecting us to our roots. A big attraction is belonging.

Another big attraction for many is having meaningful work, paid or unpaid. Most often this is through the local economy, that through their own work they create security of food, energy, water and though participation, a rich social fabric. Local economies are built around the community's purpose. Villagers/community members are employed in a range of roles. For example, hospitality, teaching, workshop delivery, gardening, local industries, feeding guests, bookings, teaching, and facility maintenance. Many use systems of work geared around the principles of trust, collaboration and felt fair decisions.

Non-violent communication, conflict management and governance are skills many communities invest heavily in. Local currencies are often used. Some of the work practices are so effective, it would make the eyes of business leaders water.

This is empowering. This is the spirit that infuses GEN and it is found to not only exist in ecovillages, but increasingly in cities, towns, backyards or balconies.

I would not like to give you the wrong impression that communities of the past offer some magical properties that solve everything. Most were restrictive, closed patriarchal and domineering, with little opportunity for individuals to exercise discretion and their freedom of choice. Flow was not part of the equation. That is why urbanisation has been so attractive, with promises of freedom of expression, of growth, of new jobs, of new careers, of new partnerships, of new different communities. Community is not a magic ingredient that serves to make humanity kinder. What we are seeing is that the search for community is on again as is interest in leaving city life, or reconfiguring urban living.

GEN exists to assist the birth of a regenerated world. A world of empowered citizens and communities, designing and implementing their own pathways to a sustainable future, and building bridges of hope and international solidarity. Its stated mission is to innovate, catalyse, educate and advocate in global partnership with ecovillages and all those dedicated to the shift to a regenerative world.

Next time you feel like taking a break, do a virtual workshop or webinar or when travel permits, visit an ecovillage. Treat yourself to an education of a lifetime – be surprised, delighted and sometimes enlightened.

GEN, the United Nations and Africa lights the way

Since 2000 GEN has had consultative status at the UN-Economic and Social Council (ECOSOC) commission and is represented at regular briefing sessions at UN Headquarters. GEN is also a partner of the United Nations Institute for Training and Research, UNITAR. This status gives GEN Advocacy the chance to join the work of various committees relevant to its concerns to promote sustainable communities and practices worldwide.

At COP 24 and 25, the GEN delegation showed off how ecovillages are already living laboratories of climate resilience worldwide. As a result Memorandums of Understanding with 12 African Countries for Regenerative Ecovillage Design was signed, plus Columbia in South America.

The intention is to develop over time, at scale, the Pan African Ecovillage Development Progamme. These governments are far-sighted in seeing the benefits of regenerating their rural sectors creating employment and attracting people back to the country.

Giving back work, meaning and empowerment

What they also realise is that here is a viable alternative to creating work in overcrowded cities. Ecovillages and eco-communities offer the opportunity to actualise at almost all of the work themes shown in Table 1 and across many disciplines; engineers, architects, builders, farmers, small scale entrepreneurs, hospitality, education, content designers, administrators, chefs – and on top of it, it is building self-sufficiency and resilience at a local level. It is one of the few 'industries' adding 'jobs' to a disappearing market while at the same time, lifting the happiness and wellbeing score of life. Government and business need to come together in support of community developments.

The United Nations Sustainable Development Goals (UN SDGS)

GEN links local developments and initiatives to benchmark overall progress against these goals. Ecovillages pioneer innovative solutions at the local level. For example, Solheimar, an ecovillage in Iceland, in a 2019 research report, ranked itself against international SDG statistics. Solheimar scored the highest overall, with 27 points out of 35 possible points. It was the leader for Goal 12, responsible production and consumption, as well as in Goal 15, life on land. For Goal 12, Solheimar had 2% of their overall vegetable production being wasted and a 70% recycling rate of their household waste. For Goal 15, Solheimar had 52% of their land as forest with 286 trees planted per person per year (Alonso, 2019).

This can be scaled to regions and states to achieve their UN Sustainable Development Goals and Climate Agreements. In 2017, GEN investigated the impact of 30 diverse ecovillages to find out how ecovillages are contributing to reaching the SDGs and Paris Climate Agreements. The TESS research project concluded that if only 5% of the EU were to engage in effective community-led climate change adaptation initiatives, carbon savings would be sufficient for 85% of its countries to achieve their 2020 emission reduction targets. TESS also showed that 63% of the surveyed community-based initiatives have been replicated elsewhere. The potential to scale community-led regeneration is high (6).

MEASURING THE IMPACT OF ECOVILLAGES

In 2017, GEN investigated the impact of 30 diverse ecovillages. In 2020 we will examine 100 more - to find out how ecovillages are contributing to reaching the SDGs and Paris Climate Agreements. Some of the findings from 2017:

GOAL 4 : QUALITY EDUCATION
100% provide education and lifelong learning opportunities in the fields of sustainable development, regenerative lifestyles and climate change adaptation.

GOAL 5 : GENDER EQUALITY
90% have more than 40% women in decision-making bodies.

GOAL 6: CLEAN WATER AND SANITATION
97% actively work to restore or replenish water sources and cycles.

GOAL 11: SUSTAINABLE CITIES AND COMMUNITIES
100% actively safeguard regenerative cultural traditions using local sustainable ways of building, farming and preparing food.

GOAL 12: RESPONSIBLE CONSUMPTION AND PRODUCTION
90% recycle, reuse and repair more than 50% of consumer goods. 85% compost all their food waste.

GOAL 13: CLIMATE ACTION
90% work actively to sequester carbon in soil and biomass

GOAL 15: LIFE ON LAND
97% of showcase ecovillages work actively to restore damaged or degraded ecosystems.

GOAL 16: PEACE, JUSTICE AND STRONG INSTITUTIONS
100% provide education in decision-making and mutual empowerment skills; 96% provide training in nonviolent conflict resolution, and 80% have an agreed-upon method for resolving conflicts.

GOAL 17: PARTNERSHIPS FOR THE GOALS
95% regularly engage in campaigns to protect human rights, the rights of communities and the rights of nature.

In their efforts to restore ecosystems, water cycles and the atmosphere, ecovillages use and often teach many of the top 100 carbon drawdown solutions, as identified by Project Drawdown in 'the most comprehensive plan ever proposed to reverse global warming'.

Figure 1: Measuring the impact of ecovillages in contributing to the UN SDGS.

Conclusion

There is a growing global cry for an alternative life model that increases meaningful employment, reverses unsustainable urbanisation, improves land ownership through regeneration, provides people with basic security, redresses the increasing loneliness, builds caring relationships and restores damaged, vulnerable ecosystems. We need slower, kinder, connected societies. This is a new wave we need to welcome in.

GEN supported by its ecosystem of rich networks offers such a vision. It is also a sound hedging against climate change and future security. We will watch with interest as those twelve African countries work to deliver their far-sighted visions.

We need to scale sustainability education to create a movement of billions, people with secure homes, in caring communities with meaningful work that provides security for water, food, energy and responsive and equitable local institutions in a friendly, healthy environment and connected planet. Sadly, global leaders have missed this point, it's not about a vaccine and returning to past. Covid-19 is nature's marketing campaign of what is in store if we don't transform ourselves by going back to the future.

References

Alonso, D.R. (2019). *Solheimars' contribution to the sustainable development goals of the United Nations,* Avans University of Applied Sciences.

Csikszentmihalyi, M. (1990). *Flow: the psychology of optimal experience.* New York: Harper and Row.

Craddock, K. (2014). Requisite Leadership Theory: An Annotated Research Bibliography on Elliott Jaques, Including: Requisite Organisation – The Glacier Project – Stratified Systems Theory – Levels of Mental Complexity – Complexity of Information Processing – The Quality of Labor – The Mid Life Crisis – and Psychoanalysis. (covering 1942 – 2002). Columbia University. (Accessed at February, 2019 from Global Organisation Design Society).

Harari, Y. (2014). *Sapiens: a brief history of humankind,* London: Harvill Secker.

Harari, Y. (2018). *21 Lessons for the 21st Century,* London: Jonathan Cape.

Newport, C., (2016). *Deep Work,* London:Piatkus.

Olivier, A . (2014). Study done at Findhorn Ecovillage. Unpublished. Directors and Trustees found to sit across two themes of work.

Tofler, A. (1980. *The third wave,* Bantam Books.

Online resources

1. IMF. (2020). *World Economic Outlook Report.* https://www.imf.org/en/Publications/WEO/Issues/2020/09/30/world-economic-outlook-october-2020 (Accessed December 18, 2020).
2. Barton, D., Manyika, J., Williamson, S.K., *Finally, evidence that managing for the long term pays off.* Harvard Business Review. https://hbr.org/2017/02/finally-proof-that-managing-for-the-long-term-pays-off (Accessed December 18, 2020).
3. Ecovillage Principles - https://ecovillage.org/projects/

dimensions-of-sustainability/ (Accessed December 18, 2020).

4. Roysen, R., Cruz, T.C. (2020). Educating for transition - ecovillages as classrooms. *International Journal of Sustainability in Higher Education* https://www.researchgate.net/publication/342442418_Educating_for_transitions_ecovillages_as_transdisciplinary_sustainability_classrooms (Accessed December 18, 2020).

5. Nashira Ecovillage. https://acei-global.blog/2019/04/26/nashira-eco-village-in-colombia-a-matriarchal-example-of-women-empowerment/ (Accessed December 18, 2020).

6. GEN Annual Report. (2019). https://ecovillage.org/our-annual-report-2019-is-here/

The way we see the world shapes the way we treat it. If a mountain is a deity, not a pile of ore; if a river is one of the veins of the land, not potential irrigation water; if a forest is a sacred grove, not timber; if other species are biological kin, not resources; or if the planet is our mother, not an opportunity – then we will treat each other with greater respect. Thus is the challenge, to look at the world from a different perspective.

David Suzuki
Environmental activist

Chapter 8:

Defending Small-Scale Agriculture as a Vital Part of an Ecological Civilization

Colin Tudge

Originally published as: Tudge C (2018) Lies, misconceptions and global agriculture. *The Ecological Citizen* 2, p. 77–85.

Editor's note: When considering the type of chapter that we wanted for this book on the subject of food and agriculture – that is, an honest account of the state of play and hope for the future – it turned out that such a piece had already been written. I am thus most grateful to Colin Tudge for allowing us to republish his article from *The Ecological Citizen* in order to bring it to an even greater audience. It appears here with a couple of very minor edits and under a new title.

More or less everything that we are told about food and farming by the oligarchs who dominate our lives – the government, the corporates, big finance and large, but mercifully not all, sections of academe – is untrue, or at least is seriously misleading. This is why the world is in such a mess – and why we must take matters into our own hands. The misconceptions that underpin present-day agricultural strategy reflect the over-confident, ultra-'rational', reductionist, materialist, positivist and imperialist mindset of the post-Enlightenment Western world. The general, almost unquestioned, assumption is that humanity's task in life is to make ourselves more and more comfortable; that this can be achieved only, or primarily, by producing more and more *stuff*, including food; that it is possible to go on producing more and more, even though the Earth is finite, because technology will always find a way; that, indeed, the pursuit of science will one day make us both omniscient and omnipotent, so we'll soon understand everything and be able to control everything for our own purposes; that this – essentially Western – way of thinking is superior to other ways of thinking (because those who think in the Western way become technically powerful and so are able to dominate the rest); and hence that the present world, led intellectually by the West, is on the right lines (despite appearances) and we can safely put our trust in our present leaders.

All these beliefs must be re-examined. Here, though, is a re-examination of six particular untruths that have come to dominate global agriculture and are leading the world hopelessly astray.

Untruth 1: We must produce more and more food

In 2011, the UK government told us that humanity needs to produce 50% more food by 2050 just to keep pace with rising population and rising 'demand' – especially for meat (Government Office for Science, 2011). The United Nations' Food and Agriculture Organization has long argued that we need to increase food production by 60–70% by that date (e.g. Alexandratos and Bruinsma, 2012); and a much-cited study published in the prestigious *Proceedings of the National Academy of Sciences of the United States of America* has argued that food production needs to *double* by then (Tilman et al., 2011). In short, the emphasis must continue to be on production, production and ever more production.

In truth...

According to Professor Hans Herren, President of the Millennium Institute (Washington, DC, USA), the world already produced enough food in 2011 to feed 14 billion people (Roseboro, 2011). This is nearly twice the present world population and, based on the United Nations' estimate of a world population of around 11.2 billion in 2100

(United Nations, 2017b), it is 25% more than we are likely to need at any point this century (see Kuhleman [2018] for more on food and population). The continued emphasis on production has nothing to do with real need, and everything to do with commerce.

Anyone who wants to can easily check current figures for themselves. The World Bank tells us that the world currently produces nearly 3 billion (metric) tonnes of cereal per year. Since one tonne contains enough energy *and protein* to feed more than 1,000 people for a day, our current annual global production of cereal contains enough macronutrients to feed more than 8 billion people. But cereals account for only half our food – the other half comes from pulses, nuts, tubers, fruit and vegetables, meat, dairy and fish. So the current total is enough for 16 billion-plus.[1]

At present, says the Food and Agriculture Organization of the United Nations (2017), "There is more than enough food produced in the world to feed everyone, yet 815 million people go hungry." That has everything to do with economic and political inequality and general disruption (notably war) and nothing to do with the total amount of food produced (as Amartya Sen has long argued; see Sen [1982]). The emphasis must switch from production to sustainability and resilience, and to care of the biosphere, human and animal welfare, social justice and general kindness. Industrial agriculture is anything but sustainable – it is a major cause of global warming (contributing nearly one-third of greenhouse gas emissions [Gilbert, 2012]) and the prime cause of the mass extinction that now threatens the majority of the world's species. It is certainly not kind, or just, and has little to do with human well-being. For while a billion go hungry a billion more suffer 'diseases of excess'. Among other things, the world population of people with diet-and lifestyle-related diabetes now exceeds the total population of the US by some margin (World Health Organization, 2017).

1. These estimates are very conservative. For information on global cereals production, see World Bank open data at https://is.gd/qZXUwy. A tonne of cereal contains (as a conservative estimate) at least 3 million kcal (Nelleman et al., 2009). The recommended energy intake for an adult male is 2,500 kcal per day; for an adult woman it is 2,000 kcal per day (World Health Organization, 2004). For further discussion see Cassidy et al. (2013).

Untruth 2: As people grow richer, demand for meat increases

This is obvious from the fact that as societies are 'lifted out of poverty' meat consumption rises prodigiously. For example, the US became hooked on steaks and burgers during the post-war economic boom of the 1950s and 60s; and the Chinese – for centuries sustained on rice with very little use of meat – are now 'demanding' all the pork, beef and chicken that they can produce themselves and the rest of the world can supply them with. Beijing and other big Chinese cities bristle with burger joints. In Britain, successive secretaries of state have told farmers that they should strive to produce more and more pork and beef for export to China.

In truth...

Nutritionists have been telling us for decades that we, human beings, do not *need* a great deal of meat, and of course, many people live long and agreeable lives on a vegetarian or even a vegan diet. Most people, it seems, do like meat, but there is very little evidence for active 'demand'. No-one to my knowledge has ever taken to the streets with placards demanding more meat, in the way that they have often demanded fairer wages or more jobs or rights for various minorities. The evidence, when looked at objectively, is that people eat what is available and what – for whatever reason – is deemed high status and fashionable.[2] We need not assume that the observed correlation between rising wealth and meat consumption is any more than a tautology: that meat in general is expensive and as people grow richer they can eat more expensive things – not just meat but also chocolate and cream cakes and a better class of alcohol. California and Germany are amongst the world's wealthiest regions, yet are also the world's epicentres of vegetarianism. More generally, *all* the world's greatest cuisines – from Italy to China via the Middle East and India – make only sparing use of meat, as garnish or stock and

---

2. In support of this, in a meta-analysis Kearney (2010) observes that whilst it is universally true that as poor nations become wealthier, the total energy intake of the populace tends to increase (i.e. people tend to consume more calories in total), it is not universally true that a substitution of meat for non-meat dietary sources occurs. As he notes, such substitution is "country-specific and is influenced by culture, beliefs and religious traditions."

for occasional feasts. Thus a low-meat diet does not mean austerity. We just need to re-learn how to cook.

The real reason for promoting meat so vigorously is not to meet the needs or satisfy the deepest desires of the human race. It is to dispose of arable surpluses. On industrial farms, which Western governments now put their weight and our money behind, more and more livestock is raised largely or exclusively on cereal and soya (not least on what is referred to in the US as 'CAFOs' – 'concentrated animal feeding operations'). The greatest problem for world agriculture is not to produce enough food (see my discussion of Untruth 1 above) but to avoid producing too much, for surpluses tend to be sold unprofitably or even at a loss. We already produce far more grain and other staples than the world really 'needs'; but animals can consume all the cereal (and soya) that arable farmers can produce – provided producers and processors can hype up the demand for meat. This they do; and a variety of academics and policymakers are content to put their critical faculties on hold and help them to do this.

If even the livestock market is glutted (perhaps because farmers do not have enough animals to consume all the surplus cereal and soya that is going), then these days that surplus can be turned into alcohol and called 'biofuel', of which modern governments like those of Britain and the US make a virtue, and support with public money. In other words, industrial farmers solve the problem of surplus cereal by *burning* it – profitably. This is economically ingenious, but it does humanity little or no good and does the biosphere a great deal of unnecessary harm.

Untruth 3: We need ever more productive crops and livestock

We are further assured that the huge increase in food that we allegedly need can be provided only by raising, yet further, the already prodigious output of our cereals and livestock – and this is to be achieved by ever more intensive breeding and nutrition. Thus we need wheat that yields at least 10 tonnes per hectare on average, about three times the yield of 100 years ago (the British average is already nearly nine tonnes per hectare [Department for Environment, Food and Rural Affairs, 2015: 1]). We need cattle that give at least 10,000 litres (2,000 gallons) per lactation, which basically means per year – which many do already: about six times as much as a wild cow would produce for her calf, and three times more than would have satisfied most farmers of the early 20th century. Broiler chickens are already expected to reach supermarket weight at six weeks and we need to make them even faster, bigger or both – and cheaper. Sows in Australia produce an average of 22 live piglets a year in two litters, while those in the US manage 28 – about four times the typical 'output' of wild boars (PigCHAMP, 2017).

In truth...

Given that we already produce far more food than we really need, and do not in fact require a lot of animal protein for a healthy diet, it follows that the rapid-growth chicken and the prodigiously fecund sow are simply unnecessary. So too is the 2,000-gallon-plus cow, which commonly suffers mightily from mastitis and lameness and is usually slaughtered after two or three lactations (traditional dairy cows commonly managed 10 or more). So too are 10-tonnes-per-hectare cereals, which in large part are grown to feed these beasts. Such crop yields, produced year after year with the aid of artificial fertilizers, rapidly exhaust the soil and destroy its structure. It is now reckoned that many fields in parts of the UK, for example, will not be farmable, at least for cereals, for more than another 30 years or so (Department for Environment, Food and Rural Affairs, 2018). Indeed, according to the United Nations Convention to Combat Desertification, about a third of all the world's agricultural soils are now seriously degraded – largely, and to some extent entirely, because of such intensive, industrial farming (United Nations, 2017a).

Untruth 4: Only high-tech can save us now

We are also given to understand that to go on feeding ourselves we need the highest of high technology. Meat substitutes made from soya or fungi are already commonplace but we could, we are told, bypass the need to raise whole organisms and simply culture animal cells *en masse* in the laboratory. The food industry is working on it (e.g. Fassler, 2018).

Above all, we are told, we need genetically modified organisms (GMOs), tailor-made by genetic engineering. GM soya, maize and rape (canola) is already sweeping the world. GM crops are not yet grown commercially in Britain and mainland Europe but are freely imported from the US and elsewhere – mainly for animal feed. Although there are plenty of protestors, GM maize, soya, canola and cotton are now broadly accepted in the US, for example, as the normal way of things.

We are told, of course, that GM crops can be and are bred specifically to be pest- and disease-resistant and so can outyield conventional types without the protection of pesticides. But most popular are the GM crops that are herbicide-resistant – enabling farmers to spray their fields not exactly with abandon but without too much restraint, to kill the weeds without killing the crops. The GM seeds and the herbicide are sold as a package.

In truth...

It is very hard to find any clear examples of GM crops that have been of unequivocal benefit to humankind. Almost always they serve mainly or entirely to make rich people richer (the biotech companies and big industrial farmers) but they solve no problems that really need solving, and (despite denials) are causing enormous collateral damage. But then, modern industrialised agriculture is entirely profit-driven and it is the rich who make the rules, so GMOs are becoming *de rigueur.*

In addition, the global biotech lobby is tremendously strong, its arguments are seductive and many politicians are taken in by them. Especially amenable are those with no scientific background who want to appear *avant garde,* up to date, modern and 'progressive'. To take two UK examples, Tony Blair, who read law at Oxford, was a positive GMO zealot, and so too is Lord Dick Taverne, who studied ancient history at Oxford, became a QC and founded the pro-GM charity Sense about Science in 2002. They, in common – alas! – with most scientists, seem not to realise that science has significant limitations: it cannot lead us to omniscience, and ideas that are exciting in the laboratory do not necessarily lead to good outcomes when applied in the real world.

Appropriate technology that truly makes life easier is certainly worthwhile, and some appropriate technology is indeed high-tech – like the mobile phone or solar panels. But much of today's agricultural high-tech – including the much vaunted GMOs as outlined above – is not appropriate at all: not needed, and often very damaging. There is a huge and growing literature on this not only in polemical articles but in scientific journals (not all scientists are on the side of big business). However, this literature is simply ignored in the most influential circles – or else answered with statistical quibbling, mostly of the kind that could be applied to almost any scientific study, if anyone cared to do so. The quibbling does not answer the objections, but it can hold things up and wear the opposition down – which is what it is intended for.

The biggest point perhaps is that large, high-tech, monocultural farms are *not* the most productive – certainly not when measured over longer time periods. A growing literature shows that small mixed farms, well managed, can be at least as productive in any one year as the big monocultures, and generally are more productive when measured over decades (for a summary of some of this literature, see Winter et al. [2016] and Laughton [2017]). This is because the mixture of crops and animals on such farms leads to resilience, so the mixed farms are better able to resist setbacks, such as droughts or late frosts. Mixed crops and livestock are also far more resistant to disease, and simply do not require large inputs of pesticides and antibiotics.

Untruth 5: Fewer workers means greater efficiency – and efficiency is all

Then, we are told, we need to reduce the farm labour force worldwide to make it more 'efficient'. A key measure of efficiency, after all, is output per worker. If the workforce is reduced and the remaining workers produce as much or more than before, then that is, surely, an increase in efficiency. And efficiency is good, is it not? The antithesis, after all, is *inefficiency*, which means waste – which is obviously bad, is it not? Who can doubt that?

Therefore, we are told, the whole world should strive to industrialise its farming. As far as possible we must replace farm labour (stroppy; inefficient; gets sick) with machines, industrial chemistry (including fertilizers, antibiotics, fungicides, herbicides, insecticides, acaricides and nematocides), and of course biotech. Machines do not deal easily with mixtures of crops and livestock and so, for reasons of efficiency, farming must as far as possible be monocultural – just one crop, or type of beast, at a time. All should be increased to achieve economies of scale: combine harvesters as big as small houses; trucks the size of small warships; small fields merged into bigger and bigger fields and small farms merged into vast estates. So it is that there are farms in the UK exceeding 1,000 hectare (roughly 2,500 acres) with just one full-time employee (though many rely on seasonal groups of immigrants, bussed in to do the fiddly bits). There are farms in the Ukraine bigger than Kent; there is a cattle farm in Australia larger than the state of Israel. This too is modernity.

Monocultural farms that produce vast quantities of just one thing at a time cannot of course feed local populations who do not live by chickens or maize or rapeseed alone. Instead they must treat all their crops and animals as *commodities* to be produced on the largest scale and sold into the global market where they are processed, packaged and distributed by evermore labyrinthine routes (with plenty of scope for chicanery, profiteering and general malpractice) to the far corners of the Earth.

So it is that Britain's farm labour force has dropped from around 700,000 in 1984 (when industrialisation was already well advanced) to 466,000 in February 2018 – that is, a mere 1.5% of the total UK workforce is now employed in agriculture. In the US, about 8% of the labour force was engaged in agriculture in 1960, and, as in the UK, this has now declined to about 1.5%. To take another telling statistic, in the ten years between 2006 and 2016 the number of dairy farmers in Britain fell by 50% from 21,000 to 10,500 – and still the decline continues. These trends are worldwide: agricultural workers have declined from 42% of the global labour force in 1991, to 28% in 2017. It is a fair bet that many of the 1 billion people who now live in urban slums in the Global South are dispossessed farmers or their dependants.[3]

---

3. UK figures are from the Office of National Statistics (2018). For the US, see Bureau of Labor Statistics data at https://is.gd/HdCmJf. For global statistics, see World Bank open data at https://is.gd/H0t247 (Accesssed March 2018).

All this is necessary, we are told. It keeps costs down and people above all 'demand' cheap food. Already in Britain, 1 million people a year must resort to food banks, and if we farmed any differently, it might seem, the figure would be higher.

In truth...

'Efficiency' is a horribly abused concept. If efficiency is defined in monetary terms, then it depends entirely on an economic context which, in reality, is highly contrived (no matter how often we are told that prices are determined by the dispassionate forces of the 'free market'). For example, industrialised agriculture is highly dependent on oil and is cheaper than traditional forms of agriculture only because oil is still available, for the time being, and its price is regulated to make sure it is still affordable (just). Perhaps even more to the point: industrial farming is cheap only because the collateral damage it causes is largely uncosted – including the cost of mass unemployment, in money (including international aid) and human misery, as the countryside worldwide is depopulated. These costs are not attributed to industrial farming (Fitzpatrick and Young, 2017). Nor is the cost in non-cash terms (or even in cash terms) of mass extinction. All this collateral damage is, in economic jargon, an 'externality'. Nothing to do with me, guv.

Nor, when you analyse it, is the industrially produced food sold in supermarkets anything like as cheap as it may seem to be; and nor can the full cost be laid at the feet of the farmer and agricultural labour. In truth, with the industrialised food chain the farmers typically get less than 20% of the retail price and their poor benighted employees, who are regularly thrown out in the name of 'efficiency', probably account for only 10% of the retail price (at most). The 80% that remains goes on big machines and fancy forecourts and packaging and razzmatazz and layers and layers of managers and shareholders and bankers who lend the money to make it all possible.[4] This all

4. The US Department of Agriculture estimates that, in 2017, approximately 16 cents in every dollar spent on food and drink in the US went to farmers, with off-farm costs (including marketing, processing, wholesaling, distribution and retailing) making up the remaining 84 cents (National Farmers Union, 2017).UK

contributes to GDP, but it does not contribute to human wellbeing and does enormous harm to the biosphere.

Untruth 6: Organic farming is a middle-class indulgence – strictly niche; it cannot possibly feed the world

As for organic farming – don't be ridiculous! If all the world farmed organically, food would cost a fortune and half the world would starve. Either that or we would all have to be vegans, and austere vegans at that, living on fibrous bread and lentil soup. It is true that sales of organic produce are going up globally, but they still amount to only a tiny proportion of the total spend on food and drink. For example, in the UK, organic makes up only about 1.5% of the total spend on food and drink, whilst in the USA, it amounts to only about 5% (Organic Trade Association, 2018; Soil Association, 2018). Organic food is elitist; strictly for the well-heeled, elite middle class. To recommend it for the world at large is simply to be irresponsible. Only high-tech, industrialised farming, driven by the competition of the global 'free market', can deliver on the largest possible scale.

In truth...

Organic farming, so despised by the powers that be, dismissed as an elitist myth, ticks all the boxes that really matter. Well-managed organic farms can be at least as productive as 'conventional' farms that rely on artificial fertilizers, pesticides and the rest (e.g. Badgley and Perfecto, 2007; Nemes, 2009; Seufert et al., 2012). The produce is of course free of pesticide residues and generally is high in essential vitamins and minerals. Organic farms employ more people – which in this populous world should be seen as a good thing; and with *appropriate* technology, the jobs they provide can be highly agreeable and sociable – the basis of truly fulfilling careers.

---

figures are from the Office of National Statistics (2018). For the US, see Bureau of Labor Statistics data at https://is.gd/HdCmJf. For global statistics, see World Bank open data at https://is.gd/H0t247 (Accesssed March 2018).

What is to be done?

These six untruths are a fair summary of official government policy in many countries (e.g. the UK and the US) and are what you will hear from most of the important-looking people who appear on public platforms and on TV to tell us what is what. Whether the policymakers and those who inform public opinion are themselves ill-informed, or are deliberately concealing what they know to be the truth, I do not know. I suspect it is a mixture of both. Either way it is deeply reprehensible.

All in all it has long been obvious to me and a great many other people that the oligarchs who dominate our lives have lost the plot and, quite simply, are not on our side. Successive US and British governments, in particular, over the past 35 years have seen it as their role in life not directly to meet the needs of the people but to support the corporates (and banks) that are perceived to provide the wealth that is supposedly vital for our well-being – the *sine qua non*. If and when there is any money left over we can spend some of it on the biosphere, but we cannot afford to do that until, well, we are richer than we are now (or indeed are ever likely to be). Agriculture is run on this assumption – perceived somewhat chillingly as 'a business like any other'. Housing, education, transport and healthcare are increasingly all now subject to the same neoliberal mentality. Enterprises that do not yield maximum measurable wealth in the shortest time – and concentrate that wealth so that it benefits those who do the measuring – are not considered 'realistic'. Apparently it is more important to maximise wealth, expressed as GDP, than it is to promote human well-being and to keep the biosphere in good health. However, agriculture is in fact of tremendous importance and we simply cannot afford to leave it to the present oligarchy, driven as it is by this neoliberal mindset.

The UK as an example

In the remainder of this article, I focus on the example of the UK, but I hope that it will be useful for readers in other countries facing similar issues. In Britain, the implication of my argument is that we cannot afford to leave agriculture to the Department for the Environment, Food and Rural Affairs. Indeed, the word 'agriculture' has been air-brushed out of the department's title – I suspect in anticipation of a time when British agriculture itself will be air-brushed out, like coal-mining, because Brazil and Africa have more sunshine and cheaper labour and at least for the time being can grow what we need more cheaply than we can grow it ourselves. Neither can we afford to leave agricultural science to the Biotechnology and Biological Sciences Research Council – the title of which again makes no reference to agriculture, which is now officially conceived, apparently, as a branch of biotech.

We, people at large, need to take matters into our own hands. I have summarised some of the things we need to do and can do (and, here and there, are already being done!) in *Six Steps Back to the Land* (Tudge, 2016). In particular, in Britain I suggest we need a new, quasi-independent agency, or series of agencies, to run food and farming – similar to the community-organised agencies that plan and run the dikes of Holland, without which the country would be submerged. The Dutch long ago acknowledged that the dikes were too important to be run by governments, subject to political ambition and whim. This agency for food and farming that we need must be run, not as such agencies often have been, by the great and the good and their spouses, but by people who really know what needs doing, which mainly means farmers, cooks, and conservationists, with input from scientists, sociologists and people at large who give a damn.

Two initiatives that I am heavily involved in – the Campaign for Real Farming (www.campaignforrealfarming.org) and the College for Real Farming and Food Culture (www.collegeforrealfarming.org) – are aimed at contributing not only to better food and farming but to grass-roots control dedicated to this end. *Six Steps Back to the Land* discusses ways in which people who may have never thought much about farming can get involved, and how communities can start to run things for themselves. As an important new book (Pimbert, 2018) argues, people *everywhere* must get more involved – not simply in on-the-ground farming but in shaping policy. The book is truly radical, and right now radical thinking is vital.

However, farmers cannot farm in the way the world really needs unless people at large buy their produce; sound farming depends on a sound food culture. People at large need to give a damn, and although we cannot all be farmers and do not all want to be, we can all take a serious interest in food. This means, as far as possible, buying only from growers and farmers who are doing the job in the right way. Vitally, too, we must re-learn how to cook. Governments that encouraged this really would be doing something useful.

It is not quite too late to bring the world out of its tailspin but only we can do it. Governments and big industry and the world's most powerful financiers are looking the other way.

References

Alexandratos. N. & Bruinsma, J. (2012). *World Agriculture Towards 2030/2050: The 2012 revision* (ESA Working Paper 12 – 03). Food and Agriculture Organization of the United Nations, Rome, Italy. Available at https://is.gd/EWc2Ur (Accessed March 2018).

Badgley, C. & Perfecto, I. (2007). Can organic agriculture feed the world? *Renewable Agriculture and Food Systems 22,* p. 80-5.

Cassidy, E., West, P., Gerber, J. & Foley, J. (2013). Redefining agricultural yields: from tonnes to people nourished per hectare. *Environmental Research Letters 8,* p. 034015.

DEFRA (Department for Environment, Food and Rural Affairs). (2015). *Farming Statistics: Provisional 2015 cereal and oilseed rape production estimates, United Kingdom.* London. https://is.gd/ezdryA (Accessed March 2018).

DEFRA. (2018). *The future farming and environment evidence compendium.* London. https://is.gd/kaq2rJ (accessed March 2018).

FAO (Food and Agriculture Organization of the United Nations). (2017) *How close are we to #ZeroHunger?* Rome. https://is.gd/OPHPqp (Accessed March 2018).

Fassler, J. (2018). What the alt-protein revolution tells us about the future of eating. *The New Food Economy.* https://is.gd/8ckeDy (Accessed March 2018).

Fitzpatrick, I. & Young, R. (2017). *The Hidden Cost of UK Food.* Sustainable Food Trust, Bristol. https://is.gd/iKoHqI (Accessed March 2018).

Gilbert. N. (2012). One-third of our greenhouse gas emissions come from agriculture. *Nature News*. Available at https://is.gd/RYxOsU (Accessed March 2018).

Government Office for Science. (2011). *The Future of Food and Farming: Final project report*. London. https://is.gd/48CcxZ (Accessed March 2018).

Kearney, J. (2010). Food consumption trends and drivers. *Philosophical Transactions of the Royal Society B 365*, p. 2793-807.

Kuhlemann, K. (2018). 'Any size population will do?': The fallacy of aiming for stabilization of human numbers. *The Ecological Citizen 1*, p: 181-9.

Laughton, R. (2017). *A Matter of Scale: A study of the productivity, financial viability and multifunctional benefits of small farms (20 ha and less)*. Landworkers' Alliance and Centre for Agroecology, Coventry, UK. https://is.gd/l5euEL (Accessed March 2018).

National Farmers Union. (2017). *The Farmer's Share*. https://nfu.org/farmers-share/ (Accessed March 2018).

Nelleman, C., MacDevette, M., Manders, T., et al. (2009). *The Environmental Food Crisis. United Nations Environment Programme*, Nairobi. https://is.gd/znLneu (Accessed March 2018).

Nemes, N. (2009). *Comparative Analysis of Organic and Non-Organic Farming Systems: A critical assessment of farm profitability*. Food and Agriculture Organization of the United Nations, Rome. https://is.gd/thlVMO (Accessed March 2018).

Office for National Statistics. (2018). *Labour in the Agriculture Industry*, UK. https://is.gd/SqCI3M (Accessed March 2018).

Organic Trade Association. (2018). *Organic Market Analysis*. https://is.gd/wWIso2 (Accessed March 2018).

PigCHAMP. (2017). *Benchmarking Summaries*. https://is.gd/PqHXkp (Accessed March 2018).

Pimbert, M., ed. (2018) *Food Sovereignty, Agroecology and Biocultural Diversity: Constructing and contesting knowledge*. London: Routledge.

Roseboro, K. (2011). Leading scientist says agroecology is the only way to feed the world. *The Organic and Non-GMO Report*. https://is.gd/178VOY (Accessed March 2018).

Sen, A. (1982). *Poverty and Famines: An essay on entitlement and deprivation*. Oxford: Oxford University Press.

Seufert, V., Ramankutty, N. & Foley, J. (2012). Comparing the yields of organic and conventional agriculture. *Nature 485,* p: 229-32.

Soil Association. (2018). *The 2018 Organic Market Report*. https://is.gd/lR1azE (Accessed March 2018).

Tilman, D., Balzeri C., Hill, J. & Befort, B. (2011). Global food demand and the sustainable intensification of agriculture. *Proceedings of the National Academy of Sciences of the United States of America 108, p.* 20260-4.

Tudge, C. (2016). *Six Steps Back to the Land*. Cambridge: Green Books.

United Nations (2017a) *Global Land Outlook*. United Nations Convention to Combat Desertification, Bonn. https://is.gd/93M2v4 (Accessed March 2018).

United Nations (2017b) *World Population Forecasts: The 2017 revision*. United Nations Department of Economic and Social Affairs, New York, NY. https://is.gd/mPaOt7 (Accessed March 2018).

Winter, M., Lobley, M., Chiswell, H. et al. (2016). *Is there a Future for the Small Family Farm in the UK?* Prince's Countryside Fund, Londonhttps://is.gd/NzdgWh (Accessed March 2018).

World Health Organization. (2004). *Human Energy Requirements*. Rome. https://is.gd/ZtTYBF (Accessed March 2018).

World Health Organization (2017) *Diabetes: Fact sheet*. https://is.gd/YitCkV (Accessed March 2018).

Chapter 9:

Covid-19 Disruption of Schooling and Radical Reform in Education

Peter Gray

A pandemic is a terrible thing. By definition, it causes massive illness and death. It also causes economic devastation for many, and it exacerbates social inequality, as those who are poor to begin with suffer more than the rich. This is true of past pandemics as well as the current Covid-19 pandemic. But pandemics can also have silver linings, as they can shake people out of old ways of thinking and doing and set the course for new, improved ways. For example, historians have contended that the 14th century Black Death pandemic disrupted stagnant medieval beliefs, catalysed large-scale social reforms, and helped bring on the Renaissance period of renewed humanism and interest in learning (Griffin & Denholm, 2020). More recently, and on a shorter time scale, the HIV/AIDS pandemic, which began in the 1980s, brought many gays out of the closet, which led many people, of all political and religious persuasions, to realise that some of their dear friends and relatives were gay, which may thereby have prompted the subsequent rapid revolution in social acceptance of gays and lesbians.

Might there be a silver lining to the present Covid-19 pandemic in the realm of education? That is the question of this chapter. To begin I must distinguish between two terms that appear in the chapter's title, *schooling* and *education*. These are commonly used as synonyms, but in my lexicon they are not. *Schooling*, as I use the term, refers to sets of procedures employed by specialists, called teachers, to induce people, called students, to acquire a specified set of skills, knowledge, values, beliefs, and/or ideas, referred to as a curriculum. *Education*, as I use the term, refers to a much broader concept. It can be defined as everything a person learns that helps that person to live a satisfying, meaningful, and moral life. By this definition, education does not include everything a person learns, but includes everything a person learns that is, in the long run, helpful to that person and to the society in which the person lives. Most of education, by this definition, occurs outside of schooling. Schooling can contribute to a person's education, but it can also detract from it.

I've divided this chapter into four sections. The first describes harmful effects of our standard system of schooling that were occurring prior to the pandemic. The second describes how children and parents coped with the immediate effects of the pandemic. The third describes how the pandemic influenced families' thinking and planning relevant to their children's education. And the fourth presents a vision for the future of education, which might be hastened by effects of the pandemic.

Pre-Covid-19 problems with compulsory schooling

In the first period of the Covid-19 pandemic, with schooling disrupted, people talked about "normal" schooling as that which was occurring before the pandemic. But that "normal" was not healthy and would not have been seen as normal decades earlier. The data I summarise here pertain to the United States, but much the same is true throughout much of the world.

Over the decades, from roughly 1960 to 2020, the amount of children's time consumed by schooling increased greatly. The increase was gradual, slow enough not to be noticed from year to year, but over the 60-year period it was huge. The average length of the school year in the United States increased by 5 full weeks, and the average length of the school day increased from 6 hours to closer to 7 (Column Five, 2020). But the biggest changes were in curriculum and homework. Recesses for elementary students were reduced or even eliminated, lunch hour became much less than an hour and was increasingly regimented, and art and music classes and creative activities in other classes were reduced or eliminated, all to permit more time to drill for the ever-growing number of standardised tests. Homework was greatly increased and began to be demanded even of kindergarten children. One study revealed that the average amount of time that schoolchildren, all grades combined, spent at school plus schoolwork at home increased by 7.5 hours per week just between 1981 and 2003, which is the equivalent of adding nearly an entire adult workday to children's weekly schoolwork (Juster et al., 2004).

By the beginning of the twentieth century, many children were spending more time at schoolwork than their parents were at full-time jobs. Children who could not adapt to all this micromanaged, often boring, sedentary activity were increasingly labelled as having one or more mental disorders, with ADHD being the most common. Children's time outside of schoolwork also became increasingly adult-controlled and school-like. Where children once roamed and played freely outdoors, they now were commonly carted from one adult-controlled sport or other adult-managed activity to another. There is no evidence that this extra schoolwork, adult regulation, and pressure increased real learning, learning that lasts beyond the next test, but there is lots of evidence that it had profound negative effects on children's mental health.

Analyses of standardised clinical questionnaires, given in unchanged form to normative groups of young people over the decades, revealed that the rates of what today would be called Major Depressive Disorder and Generalised Anxiety Disorder among adolescents increased roughly 5- to 8-fold between the 1950s and the year 2000 (Gray, 2011; Twenge et al., 2010), and other measures indicate that young people's anxiety and depression have continued to increase since then. Data collected by the US Centers for Disease Control reveal that the suicide rate for children under age 15 increased by 6-fold between the 1950s and the second decade of the twenty-first century. Other research reveals a continuous decline in young people's sense of control over their own lives over this same period, as assessed by a standard measure of locus of control (Twenge et al., 2004). And still other research reveals a continuous and overall large decline in creative thinking, as assessed by the well-validated Torrance Tests of Creative Thinking, among schoolchildren at all grade levels, at least from the mid 1980s through the first decade of the twenty-first century (Kim, 2011). None of these findings should surprise us. Take away free play and time for creative activities, reduce children's opportunities to control their own lives, and increase the stress of schooling and what do you get? Depression, anxiety, loss of internal locus of control, and decline in creativity.

Evidence for the role of schooling in the declining mental health of young people comes not just from correlations over decades, but also from correlations within the calendar year. A study published under the title *Stress in America* by the American Psychological Association (2014), found that teenagers in school were the most stressed-out people in the United States and that 83% of them attributed their stress at least partly if not fully to school. No other source of stress was mentioned anywhere nearly as often. Moreover, when the survey was conducted during summer vacation from school, the percentage reporting severe stress was half of that found when school was in session. Other research reveals that, for young people of school age, but for nobody else, the rates of emergency mental health admissions, attempted suicides, and actual suicides are roughly twice as high during weeks when school is in session compared to when school is not in session (Hansen & Lang, 2011; Lueck et al., 2015; Plemmons et al., 2018). A study of hundreds of middle-school children from many different school districts, which involved reporting on their moods at random times when a beeper went off, revealed that school was where the children were least often happy (Csikszentmihalyi & Hunter, 2003). In similar study with high-school students, 75% of the reports in school were of negative feelings, the most common of which were *tired, stressed,* and *bored* (Moeller et al., 2020).

Peter Gray

Children and teens coped remarkably well with the first Covid-19 lockdown

Most schools in the United States closed, because of the pandemic, around the middle of March of 2020, as did most sporting programmes and other formal after-school and weekend activities for children. Almost immediately, dire predictions appeared in the popular media about the harmful consequences that school closure would have for children and families. Without the structure of school and other adult-directed activities, what would children do? How could parents deal with bored, restless children at home all day? What would happen to children's minds and bodies? Would they just vegetate?

To address such questions, the nonprofit organisation Let Grow conducted a survey of parents and school-aged children in April of 2020, about a month after most schools closed, and then repeated the survey, with a new sample of participants, in May (Gray, 2020). The survey samples came from a demographically representative list of people in the United States willing to fill out survey forms, maintained by a market research company. In each month, 800 parents of children from age 8 through 13 and 800 children in that age range were surveyed. The results revealed that, all in all, families were coping very well, in some ways better than they had before schools closured. For example, approximately 50% of the children, both months, reported that they were "more calm" since the pandemic than they had been before, and the remainder were roughly evenly split between "less calm" and about the same as before. Likewise, many more parents, both months, reported that their children were "less stressed" after schools closed than reported the opposite.

The surveys also revealed that the children were getting more sleep than before. Most were doing school-imposed distance lessons at home, but a question about this revealed that, on average, it took only 3 hours per school day to complete them. This reduction in school time, combined with the closure of after-school programs, provided children with much free time. The survey indicated that most children were using that time productively to pursue activities that they did not have time for before. Among those frequently mentioned, by the parents in response to an open-ended question, were learning to ride a bicycle, exploring nature, reading for pleasure, learning new games, drawing or painting, knitting or other crafts, learning to play a musical instrument, starting to learn a new language, cooking, doing laundry, and engaging younger siblings in constructive ways. Many parents seemed to discover, for the first time, that their children thrived when they were not kept constantly busy with adult-imposed activities. In April, 73% of the parents agreed with the statement, "I am gaining a new appreciation of my child's capabilities," and only 5% disagreed (the rest were neutral). Also, in both months, most parents reported that conflicts between them and their child had *decreased* since the pandemic. A possible explanation for this is that many conflicts, prior to the pandemic, had to do with school and school-like activities—getting the kids up for school, getting them to do their homework, getting them to their various after-school activities, and dealing with children's frustrations about school.

At least two other surveys, one in the UK (Widnall et al., 2020) and one by the Wheatly Institute in the US (Twenge et al., 2020), have likewise shown that students' mental health improved during the school lockdown in the spring and early summer of 2020. The UK study showed a reduction in anxiety and the Wheatly study showed a reduction in depression. None of this is to suggest that children and teens were oblivious to the negative, frightening effects of the pandemic. The UK study revealed that they were quite concerned about the disease and the economic devastation, especially when these were affecting their own family or people they knew. Yet, overall, effects of these concerns on their mental health were apparently outweighed by the benefits of free time, family time, sleeping time, time to pursue their own interests, and the temporary reduction in the demands of school. For some families, at least, this observation has been a wakeup call.

Peter Gray

Effects of the pandemic on families' schooling plans: homeschooling, pods, & microschools

Prior to the pandemic, a small but ever-growing number of families were removing their children from public schools for homeschooling. According to surveys conducted by the United States Department of Education, the percentage of school-aged children who were homeschooling increased from 1.7% in 2000 to 3.3% in 2016 (Snyder et al., 2019), and a Gallop Poll in August, 2019, estimated that 5% would be homeschooling that fall. A year later, in August of 2020, the summer of the pandemic, a repeat Gallop Poll, with the same questions, revealed a sudden doubling of those planning to homeschool—up to 10%. What had been a gradual upward slope became a spike. The poll was clear in identifying homeschooling as "not enrolled in a formal school but taught at home." Online learning, at home, while enrolled at a school, was not counted as homeschooling.

One reason for the homeschooling surge, no doubt, had to do with uncertainty as to when and how public schools would reopen. Plans kept changing as Covid-19 cases in any given city or state fell or rose. Would school be entirely remote, with teachers in control but the children home on their computers? Or would students go to school, either every day or some days, with varying uncertain measures to reduce the risk of Covid-19? Such uncertainty led many families to see homeschooling as a better option, because it would allow them to settle on a definite plan that they could control.

For some, however, the decision to homeschool was also stimulated by the parents' observations of how well their children were faring outside of school. They saw the reduced stress and increased self-initiative in their children and began to value the increased family freedom that came with release from school schedules. Some parents reported that seeing the lessons the school was presenting to their children over the internet contributed to their decision to homeschool. Some saw how arbitrary and boring the lessons were and felt encouraged that the family could develop a more meaningful curriculum, with the aid of the internet and other resources. Some, impressed by how well their children learned on their own initiative, chose not just to homeschool, but to homeschool by the method commonly called unschooling, in which there is no imposed curriculum and children learn by pursuing their own interests. Here is what one mother wrote, in a blog comment, regarding this decision:

"I have 4 kids who have always been in the public-school system, and we felt a lot of stress about school over the years. Over the summer I joined a homeschooling Facebook group to learn more about homeschooling 'just in case' virtual school didn't work out. A couple weeks into the virtual school year, and I knew it wasn't going to work for my kids. I had quit my part-time job to help my kids with school, and I felt like the virtual program was wasting our time and adding layers of stress and boredom. I heard about unschooling on the Facebook group, and I decided to learn more… [The books I read] really opened my eyes to the problems imposed by the American public-school system and the possibility of real joy and ease through self-directed learning. We are now 3 weeks into unschooling. Playing, talking, exploring, reading, baking, video gaming, hiking, relaxing, and really enjoying each other with no schoolwork power struggles or stress. Covid-19 is horrible, but I am grateful that it has forced me to really stop and think and find a better way for my family."

With more homeschoolers come increased opportunities for homeschooling families to join together and form educational co-ops, microschools, or learning pods, which bring children together for joint learning and recreational opportunities. This trend had preceded the pandemic but spiked in the fall of 2020. Such cooperation among homeschooling families not only enriches the children's experiences, but also enables parents to share the tasks of facilitating the children's education, or, in some cases, to pool their money and hire a teacher or facilitator. These learning groups can rise quickly, can meet in living rooms or community spaces, and can be low cost or even no cost. Across the US, thousands of such learning groups were formed near the beginning of the 2020-2021 school year (Ark, 2020). It seems unlikely that this trend will reverse itself once the pandemic is over. Families that discover the increased learning flexibility, decreased stress, and increased control over their own schedules that comes with these alternatives to conventional schooling will have little incentive to turn back.

All this illustrates, for the Covid-19 pandemic, the kind of phenomenon that has been reported to occur in past pandemics. Ongoing, overarching systems break down, which leads people who had not done so before to take matters into their own hands. Sometimes what they come up with is better than what had been normative before.

A vision for the future of education

Of course, I have no crystal ball, and some might attribute the vision I describe here to wishful thinking more than to prognosticating power. However, even before the pandemic, movements toward what I describe here were beginning to occur, and these now appear to be accelerating. In brief, I envision a rational system of education, which meets the needs of our time and would be available to all, that operates in three phases over the duration of childhood and youth. The first phase involves learning about yourself, your interests, and the world around you; the second involves exploring possible career paths; and the third involves specialised professional training and certification for careers that legitimately require them. In what follows I elaborate a bit on each of these and comment on the degree to which they are already emerging. Their emergence is not the result of top-down dictates but is driven bottom-up by those who opt out of current ways and develop new ones.

Phase I: Learning about one's world, one's self, and how the two fit together

In this vision, the first 15 to 18 years of one's life are years of self-governed play and exploration, by which learners make sense of the world around them, try out various ways of being in that world, discover and pursue activities that most interest them, and create a tentative plan about how they might support themselves as independent adults. This is what happens already for the growing number of young people whose families have opted for *Self-Directed Education*, whether through home-based unschooling or enrollment in a school or learning center designed to support autonomous learning. (Note: consistent with the terminology developed by the Alliance for Self-Directed Education, I capitalise Self-Directed Education when it refers to the deliberate practice of opting out of coercive schooling and taking responsibility for one's own education.) In my book *Free to Learn* and in various academic articles, I have described how children are biologically designed to educate themselves through exploring, playing, and pursuing their own interests and have summarised research evidence that people who took this approach are doing very well in adulthood (Gray, 2013, 2016, 2017).

Self-Directed Education has never been easier to pursue than it is today—because of increasing social acceptance of it and technology that makes information readily available—and it will continue to get even easier. We also now live in a world where the rote procedures and memorised knowledge taught in schools are less useful than they were in the past. We now have robots to do rote work and search engines to store knowledge in easily retrievable fashion. What we need are people who are creative, critical thinkers, socially competent, passionate about their chosen career, and able to learn on the job. These are the traits employers are seeking. They are also the traits that our schools suppress and are most fully developed when young people take charge of their own education. The pandemic-induced school closures have proven to many families that their children can learn well without school, which has led many to homeschooling and may eventually lead many to Self-Directed Education.

Increased numbers of families adopting Self-Directed Education will result in increased public pressure to divert some of the tax money currently spent on standard schools to support learning centres and other resources for Self-Directed Education. One possibility is that public libraries will morph into such learning centres. To some degree, this is already happening as many libraries are developing maker spaces, opportunities for free play, discussion groups of all sorts, and other opportunities for learning beyond just books. As more resources become available for Self-Directed Education, more families will choose it, which will result in still more resources, in a beneficent cycle.

Phase II: Exploring career paths

For several decades now, the most common step after secondary school, for those who can afford it, is enrollment in a four-year college. Indeed, because of family and societal pressure, many young people see college as compulsory, essentially a continuation of high school—grades 13, 14, 15, and 16. Those years of schooling are even more expensive than the earlier ones, and this expense must generally be paid by the parents or through loans that can saddle a person for decades.

What does one get for that money and those additional four years of courses? One gets a bachelor's diploma, which, as our society currently operates, is a prerequisite for certain kinds of jobs. The diploma supposedly signifies that the person has been "educated" to a higher degree than someone without that diploma. But evidence is growing that little education actually occurs in those years (Arum & Roksa, 2011). Students study for tests and then forget what they learned, much as they did in high school. Students commonly choose courses because they are deemed to be easy, or likely to improve their grade-point average, or fit their preferred schedule, not because of a passionate interest. Here is how one college professor, Shamus Khan, has described this situation (Hayden, 2011): "I am part of a great credentialing mill... Colleges admit already advantaged Americans. They don't ask them to do much or learn much. At the end of four years, we give them a certificate. That certificate entitles them to higher earnings. Schools help obscure the aristocratic quality of American life. They do so by converting birthrights (which we all think are unfair) into credentials (which have the appearance of merit)."

College administrators have long argued that the main educational benefit of college is a gain in critical thinking, but systematic studies show that such gains are actually small overall, and for approximately 45% of students are non-existent (Arum & Roksa, 2011). I have been unable to find any evidence that critical thinking improves over four years of college more than it would have, in the same or similar people, if they had spent those years doing something else. In a survey conducted by PayScale Inc. a few years ago, 50% of employers complained that the college graduates they hire aren't ready for the workplace, and the primary reason they gave is lack of critical thinking skills (Belkin, 2017). My own observations suggest that critical thinking grows primarily through pursuing one's own interests and engaging in serious, self-motivated dialogues with others who share those interests, not from standard classroom practices.

One of the many problems with our current system is that, even after 17 years of schooling including college, students have little understanding of potential careers. The only adult vocation they have witnessed directly is that of classroom teacher. A student may decide, for some reason (maybe because it sounds prestigious), to become a doctor, or a lawyer, or a scientist, or a business executive, but the student knows little about what it means to be such a thing.

In the rational system I envision, students would spend time working in real-world settings that give them an idea of what a career entails before committing themselves to that career. For example, the person interested in becoming a doctor might work in a hospital for a period of time, maybe as an orderly or a medical assistant. Maybe it would be an official apprenticeship, with a bit of relevant course work as part of it, or maybe a regular job. By this means, the person would gain a practical understanding of what it is like to be a doctor and make a realistic assessment of whether or not this would be a good path for her or him. Do I like being in hospitals and around sick people? Do I have the compassion, fortitude, and reasoning skills required to be a good doctor? If the answer is no, then it is time to try out a different career path.

The same is true for any other career. The person interested in law might work in a law office; the person interested in being a scientist might work as a lab assistant or field assistant; the person interested in becoming a business executive might work as a clerk in a business setting. In this way they would further their education and gain real-world experience while making at least some money rather than spending money. In the process, they would get to know, and be known by, professionals in the realm their interests, who could write recommendations that would help in applications for further training or advancement.

Already many companies, recognising that a typical college education doesn't prepare people well the company's work, have created apprenticeship programs and dropped the requirement of a college diploma for prospective employees. According to the US Department Labor (2020), the number of officially registered apprenticeships in the United States rose from about 350,000 in 2011 to about 640,000 in 2019. As typical examples, BMW has an apprenticeship program in Spartanburg, SC, for training engineers, and at least one commercial insurance company offers apprenticeships in claims adjustment and underwriting—jobs that formerly required a college degree.

There is reason to believe that the pandemic will hasten this movement away from college and toward apprenticeships or apprentice-like jobs. The economic fallout created by the pandemic wiped out the college savings of many families, which will lead more young people to seek other routes to their preferred careers. Moreover, when colleges opened with only or mainly distance learning in the 2020-2021 academic year, many students discovered that they or their parents would be paying tens of thousands of dollars just to watch lectures and take tests online, something they could otherwise do for free. People who had not previously thought about alternative ways of gaining a higher education began to think about them. College enrollments plummeted in the fall of 2020, just as public-school enrollments did.

In my vision, universities will not disappear, but will persist as places for government-supported science and scholarship. Young people who wish to pursue such careers would become apprentices there, working alongside of scientists and scholars in the fields of their dreams.

Phase III: Becoming credentialed for specialised work

For some work it is imperative that the people doing it know what they are doing. Those are the jobs for which specialised training, guided by experts and evaluated by rigorous testing, may be essential. Anyone needing a surgeon, dentist, lawyer, or electrician would want to be sure that the person has been credentialed and licensed through means that include proof of competence and knowledge of proven best practices. This is the only phase of the educational system where testing should be essential. Such credentialing might in some cases be part and parcel of apprenticeship systems, or in other cases occur in schools for professional training, such as vocational schools, medical schools or engineering schools. So, the young woman who has explored a medical career by working as a medical assistant might, at some point, apply to medical school. For admission, she would have to present evidence that she knows what she is getting into and has prepared herself adequately to begin such training; and then, at the end, she would have to prove competence in whatever medical specialty she had chosen.

Concluding remark

In sum, what I have described here is an educational evolution that began before the pandemic and may be hastened by it into a revolution. It will be fascinating to see, over the next few years and decades, how much of this vision comes to pass.

References

American Psychological Association. (2014). *Stress in America.* https://www.apa.org/news/press/releases/stress/2013/stress-report.pdf (Accessed September, 2020).

Ark, T.V. (2020). Microschools meet the moment. *Forbes.* https://www.forbes.com/sites/tomvanderark/2020/08/04/microschools-meet-the-moment/?sh=1e5e11ba450a (Accessed October 2020).

Arum, R. & Roksa, J. (2011). *Academically adrift: Limited learning on college campuses*. Chicago: Chicago University Press.

Belkin, D. (2017). Exclusive test data: Many colleges fail to improve critical thinking skills. *Wall Street Journal*. https://www.wsj.com/articles/exclusive-test-data-many-colleges-fail-to-improve-critical-thinking-skills-1496686662 (Accessed September 2020).

Brenan, M., (2020). K-12 parents' satisfaction with child's education slips. *Gallop*. https://news.gallup.com/poll/317852/parents-satisfaction-child-education-slips.aspx (Accessed September 2020).

Column Five Media. (2020). America's schools: 1950s vs. today. https://www.columnfivemedia.com/work-items/infographic-americas-schools-1950s-vs-today (Accessed October, 2020).

Csíkszentmihályi, M. & Hunter, J. (2003). Happiness in everyday life: The uses of experience sampling. *Journal of Happiness Studies 4*, p. 185–199.

Gray, P. (2011). The decline of play and the rise of psychopathology in childhood and adolescence. *American Journal of Play 3*, p.443-463.

Gray P (2013). Free to learn: *Why unleashing the instinct to play will make our children happier, more self-reliant, and better students for life*. New York: Basic Books.

Gray, P. (2016). Mother Nature's pedagogy: How children educate themselves. In Lees H and Noddings N eds. *Palgrave international handbook of alternative education*. London: Palgrave, p. 49-62.

Gray, P. (2017). Self-directed education - unschooling and democratic schooling. In Noblit G ed. *Oxford research encyclopedia of education*. New York: Oxford University Press.

Gray, P. (2020). Kids continued to cope well two months after schools closed. *Psychology Today, Freedom to Learn*, https://www.psychologytoday.com/us/blog/freedom-learn/202008/kids-continued-cope-well-two-months-after-schools-closed (Accessed August 2020).

Griffin, D. & Denholm, J. (2020). Lessons from 4 other global pandemics throughout history. *MedicalXpress*. https://medicalxpress.com/news/2020-04-lessons-global-pandemics-history.html (Accessed October 2020).

Hansen, B. & Lang, M. (2011). Back to school blues: Seasonality of youth suicide and the academic calendar. *Economics of Education Review 30*, p.850-851.

Hayden, E. (2011). Study says college students don't learn very much. *Atlantic.* https://www.theatlantic.com/culture/archive/2011/01/study-says-college-students-don-t-learn-very-much/342624/ (Accessed October 2020).

Juster, F.T., Ono, H. & Stafford, F.P. (2004). Changing times of American Youth: 1981-2003. *University of Michigan Child Development Supplement.* https://www.researchgate.net/publication/260403511_Changing_times_of_American_youth_1981-2003 (Accessed September 2020).

Kim, K.H. (2011). The creativity crisis: The decrease in creative thinking scores on the Torrance Tests of Creative Thinking. *Creativity Research Journal 23,* p. 285-295.

Lueck, C., et al. (2015). Do emergency pediatric psychiatric visits for danger to self or others correspond to times of school attendance? *American Journal of Emergency Medicine 33,* p. 682-684.

Moeller J., Brackett, M., Ivcevic, Z. & White, A. (2020). High school students' feelings: Discoveries from a large national survey and experience sampling study. *Learning and Instruction 66.* https://www.researchgate.net/publication/335024305_High_school_students%27_feelings_Discoveries_from_a_large_national_survey_and_an_experience_sampling_study (Accessed October 2020).

Plemmons, G., et al. (2012). Suicide deaths and nonfatal hospital admissions for deliberate self-harm in the United States. *Crisis 33,* p. 169-177.

Snyder, T.D., de Brey, C. & Dillow, S. (2019). *Digest of education statistics 2017: 53rd edition,* p. 132. Department of Education: Washington, DC.

Twenge, J.M., Coyne, S.M., Carroll, J.S. & Wilcox, W.B. (2020). Teens in quarantine: Mental health, screen time, and family connection. *Report of the Wheatly Institute for Family Studies.* https://ifstudies.org/ifs-admin/resources/final-teenquarantine2020.pdf (Accessed October 2020).

Twenge, J.M., Gentile, B., DeWall, C.N., Ma, Dç, Lacefield, K. & Schurtz, D.R. (2010). Birth cohort increase in psychopathology among young Americans, 1938–2007: A cross-temporal meta-analysis of the MMPI. *Clinical Psychology Review 30,* p. 145-154.

Twenge, J.M., Zhang, L. & Im, C. (2004). It's beyond my control: A cross-temporal meta-analysis of increasing externality in locus of control, 1960–2002. *Personality and Social Psychology Review 8,* p. 308–319.

US Department of Labor. (2020). Registered apprenticeship national results fiscal year 2019. *Report on Employment and Training.* https://www.dol.gov/agencies/eta/apprenticeship/about/statistics (Accessed October 2020).

Farm animals are far more aware and intelligent than we ever imagined and, despite having been bred as domestic slaves, they are individual beings in their own right. As such, they deserve our respect. And our help. Who will plead for them if we are silent? Thousands of people who say they 'love' animals sit down once or twice a day to enjoy the flesh of creatures who have been treated so with little respect and kindness just to make more meat.

Jane Goodall
Conservationist

Chapter 10:

Energy Transformation

Marie Claire Brisbois

Energy permeates our entire lives. While we often don't think about it directly, our lighting and heating or cooling systems take energy. Getting to and from places of work and recreation takes energy. Our phones, computers, and internet networks take energy. Making and transporting food and goods all take energy. This means that any change to our energy system reaches into every corner of our lives. It also means that, as we transition to a world where carbon free-energy is the only option, our lives are going to change.

While energy *transitions* have occurred in the past we have never before needed so desperately to transform our entire energy system. Past transitions were primarily instigated by technological change. For example, when we moved from candles to electricity for lighting, it was because of technological progress and a slow but steady social uptake of electricity. We saw similar dynamics with the move from coal to gas for home heating, or from horses to automobiles for personal transportation. In general, these kinds of transitions have historically taken between 30 and 50 years (Geels, 2014). In order to prevent catastrophic climate change, we need the transition to renewable energy to occur much faster.

The climate imperative means that the current energy transition needs to be different from what we've seen in the past. There has been technological progress, notably in the form of renewables technologies like solar, wind and tidal power. We've also seen advancements in storage technologies like batteries and hydrogen. However, technological evolution on its own is not enough to make changes on the timeline they are needed. Instead, sweeping public policy and societal behaviour changes are necessary.

Covid-19 and energy use

Change is, by definition, disruptive. This has been very evident during the Covid-19 pandemic. As our lives have been turned upside-down, so too have our energy consumption patterns. For example, lockdowns and movement restrictions mean that personal transportation patterns have shifted. For many, life "before" was defined by daily automobile or public transportation-based commutes to communal workplaces. People working white collar jobs may have been regularly flying for business meetings, sales or conferences. While those working to provide services deemed as essential have still been on the move (e.g. nurses, agricultural workers, food shop attendants), the numbers of those commuting daily has decreased drastically.

From an energy perspective, the shift in mobility activities has been profound. Many people are not using energy for personal transportation at all (although transportation of the goods they consume is another story), while many others have shifted to active forms of transportation that limit virus exposure like cycling or walking. Emissions from cars and planes have dropped dramatically with many cities seeing their cleanest air since before the industrial era. While there has been a smaller reverse trend of people with cars opting for private, virus-free transportation instead of trains or buses, the general trend has been away from more carbon-intensive activity.

As lockdowns end (and restart, and end, and restart), carbon-based transportation has again increased. However, the abrupt decrease in transportation emissions, and change in commuting behaviours has proved that it is, in fact, possible to fundamentally shift how we move around, and even whether or not we need to.

The closure of many workplaces also means that many more people are now working from home. This has marked a fundamental change in energy consumption patterns. Daytime energy consumption for technology, appliances and heating used to be centred on workplaces. Under lockdown and work-from-home conditions, many people are now heating (or cooling) their homes during the day, and consuming energy for Information and Communications Technology and home appliances. There is also a substantial energy cost to internet activity. Most computer servers are located in northern countries where temperatures are lower and the costs for cooling hard working computer equipment are therefore lower. However, the energy cost for every internet search, conference call, and livestream is still significant (Hook et al., 2020).

Responses to the pandemic have also changed energy patterns around the goods we produce and consume. Globalised manufacturing activities and supply chains have been disrupted as workers have been forced to stay home and transport hubs closed. This has led to many product shortages and also turned a spotlight on local manufacturing and production. Covid-19 has revealed weaknesses in many of our supply chains, and demonstrated just how vulnerable we are to system disruption. From an energy perspective, usage has decreased as factories stayed offline and shipping containers stayed in port. At the same time, local producers, and particularly food producers not aimed at international export, have experienced a surge in demand (Thilmany et al., 2020) with associated decreases in energy usage for long distance transportation.

While energy system changes from the pandemic have been profound, many of them reflect dynamics that have proved, against all odds, to be quite welcome. Many people, particularly those in white collar jobs, have found that they like working from home and working less. People who haven't been on a bicycle since childhood have begun cycling again – and they seem to like it. Many Western cities have introduced emergency bike lanes to keep people on the move. Since those who cycle to work are markedly happier than those who commute by car or public transport, this represents what is likely to be a lasting behaviour change – so long as emergency bike lanes remain in place and cycling remains safer (Mytton et al., 2016).

From the perspective of energy usage (although certainly not from the perspective of human or economic loss), the pandemic has demonstrated that while change is disruptive, that doesn't mean that it is necessarily bad. For those in relatively comfortable social and economic situations – and this is a big caveat – many of the changes we have been forced to make have been, at a minimum, tolerable. Cycling, working less, staying in one place, buying less stuff, and sourcing from local producers has allowed to people to realise that a less consumptive world doesn't necessarily mean a less enjoyable world. From the perspective of those trying to realise a rapid and aggressive energy transition, this is very good news.

Shining on light on dark places

The energy changes made as a result of Covid-19 restrictions may have illuminated new possibilities. However, they have also revealed just how unbalanced access to energy is. For those without comfortable income, working or attending school from home means an increased struggle to pay household bills. Even though people may not need to pay for bus tickets to commute, the utility costs to keep homes warm and computers on during the day has increased. These costs used to be borne by employers and schools but lockdowns have shifted this onto individuals, many of whom simply can't afford to pay.

For the essential workers who continued to go to work, being able to afford to use energy for private cars has become an issue of survival. Many jobs labelled "essential" during the pandemic are those with some of the lowest wages. People working in grocery stores, waste collection and food production facilities often use public transportation because they can't afford to own or drive private cars. These people have been forced to continue using public transportation, even if they are worried about virus exposure resulting from spending time in enclosed spaces with other people.

These examples demonstrate how closely energy is linked to

issues of inequality. At local levels, conditions of "energy poverty," where people are not able to afford basic energy needs, are rampant. This is widely visible across countries in Africa and Latin America. However, it is also surprisingly prevalent across supposedly economically developed countries like the United Kingdom, Canada, and the United States. Where access to energy is limited, problems related to health, education levels, and opportunity follow.

Beyond local levels, energy is deeply tied up with social and political decisions that drive who and what are considered important by governments. This includes who gets support in times of crises. Because energy is so important to the smooth operation of society, those who keep the lights on, goods on shelves, and things moving around, have a lot of political influence. Our democracies themselves evolved in step with the increasing availability of cheap, readily accessible energy, originally in the form of first coal, and then oil and gas (Mitchell, 2011). This means that those that control energy resources like oil and gas companies, electricity generators and utilities, and other energy interests usually have an inside line to high level decision-makers. At the very least, these groups usually have enough money to pay for the best lobbyists that money can buy.

In many cases, we *want* energy interests to be in regular contact with governments. For example, the utilities that control the electricity grid need to be able to feed back on what's working, what's not, and where there might be problems in the future. However, when corporate energy interests lobby governments, it is almost always backgrounded by an interest in conditions that will increase their profitability. While providing energy-related services is the reason those companies exist, the focus on profits as an overarching, driving motivation often leads to perverse policy requests. For example, governments have spent millions of pounds of public money on carbon capture and storage (CCS) technologies that, if successful, would allow continued, highly profitable fossil fuel use. Despite decades of research, CCS research has failed to deliver economically viable climate solutions (Salvi and Jindal, 2019). These technologies often have such high energy needs that it ends up costing more energy (and money) to sequester carbon than you get from burning the original resource in the first place. Meanwhile, the money that has gone into CCS is money that hasn't been spent on developing other climate solutions.

These perverse dynamics aren't unique to the energy sector. Economists, global leaders and community activists are increasingly arguing that a society that mindlessly pursues profits at the expense of the health and wellbeing of people and the planet is simply unworkable. Oil and gas companies have spent decades working to deny climate change. They have halted, or watered down, life-saving climate legislation at the UN, and in countries around the world, in order to ensure increasing revenue streams. These companies also have their fingers in social policy. For example, there are places where oil and gas companies provide funding to police departments in communities where opposition to energy infrastructure is high (1).

While modern society has benefited tremendously from easily available access to energy, those benefits have been unequal. The political and economic power held by existing energy companies is largely used to support the status quo, and make sure that we use a lot of energy in forms off of which it is easy to make money. Energy interests currently hold the world back from making the changes needed to realise a fairer, more sustainable future. However, thanks to technological and social developments, the status quo is no longer our only option.

A different world is possible

Those who understand both the climate imperative and energy transitions know that minor shifts in usage will not be enough to save humanity. For example, shifting from petrol cars to electric cars may be a necessary step but widespread uptake of electric cars is not the answer. The resource demands of individual car ownership go beyond just fuel consumption. They include the metals and minerals required for car bodies, the energy needed to transport and manufacture vehicles, and energy and resource inputs needed to build and maintain widespread, high-usage road networks. This also diverts valuable space and public resources from the kinds of transportation infrastructure that are needed – installations like rail networks, walking paths, and cycle lanes. What is needed, on top of the transition to renewables and energy efficient practices and behaviours, is an overall, wholesale reduction in energy consumption. For years, those focused on energy profits have argued that a reduction in energy usage would be equivalent to returning to a pre-industrial, agrarian society. The pandemic, at great human cost, has provided initial evidence that a lower energy future can be both progressive and desirable.

There are a lot of different visions about what a different world could look like. Some of the most familiar are based on post-apocalyptic scenarios and focus on the breakdown of centralised electricity supplies, ICT systems, and a general collapse of modern society. As a society, we love the sensationalism of hypothetical (or safely removed) disaster. However, it turns out that there are a lot of stories about how we can use energy differently that don't involve societal collapse. Some of them focus just on the specifics on energy supplies. For example, big companies promote "hydrogen powered futures" where massive offshore wind farms create energy that is stored in hydrogen gas that is used much as natural gas is now. Technocratic stories like this address, to some extent, the need to shift away from high carbon energy. They are also the stories we most often hear on television and in the media because they are actively promoted by those with money to get their ideas out into the public sphere through traditional channels. These solutions, possibly at their peril, assume that the public will accept the massive infrastructure changes that will be required. They also miss out on the opportunity to address the vast web of social and environmental problems that are tied up in private control of energy sources, and energy overconsumption.

More compelling are the energy visions that weave together the need to shift energy sources and usage with solutions to other problems like public health crises, social isolation, over consumption, and inequality. These are solutions that "multisolve" or address many problems at the same time (2). The following stories weave together idealised visions of how life could be lived differently. For the most part, they are not wild and unlikely. Many of these dynamics draw upon ways of living that are quite normal for many people. All of them are possible with the technology that we already have developed, although much of it would need to be scaled up and out.

A different life

Imagine waking up and sleepily rolling over. The room is a comfortable temperature because the electric heat pump kicked in a couple of hours earlier. Your landlord had the heat pump installed a few years ago. It was part of a larger home renovation, subsidised by a government grant, and backed by new rental regulations, that put in efficient, double glazed windows, wall and roof insulation, and a smart meter. They'd done it all at the same time that they had been redoing the kitchen anyways to fix some water damage and damp issues. You were also able to take advantage of a grant to upgrade to a smart refrigerator. The washing machine was next but, in the spirit of reduce-reuse-recycle, you're waiting until it gives up for good before taking it to the local repair space to be fixed up and digitalised.

You'd honestly been a bit nervous about the smart appliances. For a long time, there hadn't really been good reasons to trust that private data would be respected and protected. However, strong public rules and regulations established a data commons and were now *de rigueur.* These rules prohibited the use of personal data for private gain. The big corporate tech monoliths of the previous era weren't allowed anywhere near personal energy data. It had been a big but important change when society had recognised that those kinds of data were too valuable to trust to any company that prioritised profits over the wellbeing of people. It was really only when trust in data systems had been restored that most people had signed up for smart metering and appliance systems. Digitalised energy had huge potential but people, quite rationally, really hadn't felt comfortable with the automated use of energy data until profiteering was taken out of the equation.

With the digitalised system, it was possible to have the heat pump turn on when both demand and prices are low, instead of having everyone set them to go on at the same time each morning. The same is true for your fridge – it can stand short interruptions in cooling activity and it makes sense to do most of the cooling when energy is abundant and cheap (1). The solar panels on your roof can provide direct household energy. You sometimes use this directly, and sometimes use it to charge up your on-site storage batteries. This doesn't require active choices but is instead automated, much as the internet is automated, by signals on energy supply and demand, as well as price. You're still connected to the grid. It's good to have the back-up and also nice to be able to make a bit of money when you're overproducing. It's not much but it offsets the cost of the electricity you do need to draw from the grid, and keeps generation profits in your hands instead of those of distant corporate shareholders.

The new type of home system you have is known as a "sunflower" home (Aronoff et al., 2019). It used to be the stuff of fantasy but they're everywhere now. It just doesn't make sense not to have a hyper-efficient and on-site generating home. With a little bit of government nudging, even the landlords got on board. To be honest, it was actually quite a lot of nudging but regulation, subsidy schemes and taxes on excess wealth generation meant that the landlords got there eventually.

Buttoning up your shirt, you head down to find breakfast. Your job means that you're one of the people who needs to actually go in to work every day. A lot of other people work from home at least a couple days a week. People work way less too. The move to a four day work week had actually started long before the pandemic. Companies had been trialling it all over the world and finding that people, especially those engaged in office work, tended to get just as much done in four days as they did in five. They just ended up being more efficient because they were better rested and more motivated – and way happier. Working less had been necessary in order to meet carbon targets (3). It had also proved to be extremely popular once people realised that they would be paid the same as before.

For many of your friends, working from home part time helped to save on transportation costs. People still went in to the office. It was possible to do a lot of things online, but there really wasn't a replacement for in-person interaction. However, office days were a bit flexible because everyone tried to time them for days when there was an abundance of solar and wind power for the people who needed to use transit to get in to the office. It had helped to realise that deciding daily activities based on the weather wasn't something new. It was something farmers, gardeners and other outdoor workers had been doing since time immemorial. Once you got into the rhythm of living a bit closer to the natural world, it was hard to imagine ever having been so disconnected.

That connection to nature was also helped by the fact that you, and a lot of other people, now regularly cycle or walk to work. People had complained about using road space for cycle lanes at first. But the networks of safe, separated cycle lanes actually made it possible to take kids to school by bike, to pack family groceries into panniers or cargo bikes, and even to carry infants around, without fearing for your life. It was something that people in places like the Netherlands and Denmark had been doing for ages but had only recently filtered to the rest of the world.

As e-bikes became cheaper for people who live on hills, or elderly people who like having the extra bit of back up power, the cycle lanes had to get better just to accommodate all the people! For the disabled people, heavy delivery vans, and emergency services vehicles that still regularly used cars, there was far less traffic and it was much easier to get around.

You'd been a bit nervous about cycling in the rain but, with the right clothes and shower facilities at work, it turned out to be actually pretty enjoyable. It was certainly better than the days when you used to spend what felt like hours stuck in traffic just to get home. There are days, however, where you do just take the bus. There are limits. The lanes are ploughed in the winter though and, even though everyone gets a bit down in the winter, the endorphins from cycling and the daily dose of sunlight and fresh air means that you just feel better about life these days.

It's still easy to remember the old land use planning paradigms where developers put in suburban housing without integrating shops and offices. In those days, people would go from their homes to their cars to their offices and then back again. Shopping was often done all at once in big shops, or online. People had gotten more and more isolated until things had reached such a desperate situation that the UK had actually implemented a Minister for Loneliness.

The planning rethink had done more than just reduce energy use by changing the distances that people had to travel for work and play. It meant that people were getting regular exercise and much more regular social interaction, just because they were out in public spaces. Housing spaces were designed to have large communal areas for leisure and play. Those changes had done more to reduce social isolation than the initial top down ministerial attempts.

Once prices of goods had been revised to reflect the true energy costs of production, shipping and disposal, well-made local goods had also become much more attractive. That meant that people spent a lot more time in local shops, supporting local economies, than they ever did before. You can still remember when people used to pour money into online shopping for cheap, disposable goods. It seemed so logical at the time but, in hindsight, there was no way that system of production and consumption could ever go on for long.

Changing those patterns of production and consumption had actually really kicked off with the changes in energy usage. In the early days, people used to call it "energy democracy." Energy democracy was a whole bundle of ideas that included citizen ownership and production of energy, creation of local energy jobs, community building, and political empowerment (Szulecki, 2018). It had begun slowly with just a few people working together to put in collectively owned wind turbines and solar panels. People had also started putting panels on their own homes. Local authorities had started to realise that they could control some of the energy system too, and use the energy they made and saved (through efficiency measures) to increase the quality of life in their area. The local authorities had started out mostly putting their efforts into social housing. Once they had managed to start pulling people out of poverty by providing warm, clean and efficient housing, they'd expanded to other areas. Now, almost half of the energy used for electricity and heating is owned by individuals, co-operatives and collectives, and local authorities.

Even while it wasn't the original intention, local and citizen ownership had actually facilitated the energy transformation. It turns out that, the more people are involved in generating and storing their own energy, the more accepting they are of new generation infrastructure, and of making the behaviour changes needed to reduce energy use (Baldwin, 2020). All the public energy efficiency campaigns asking people to use less energy for the environment hadn't done nearly as much as just putting resource control in the hands of the people who needed to make changes.

The shift in energy ownership had changed a lot of other things too. The business models these people and groups were using meant that, even though money was important, they weren't pursuing profit at the expense of social or environmental goals. These energy owners were able to operate for the benefit of people and the world, not just to extract profits for shareholders. The local characteristic also meant that people had to live with the consequences of their business decisions. It turns out that people are much less likely to want to put in polluting or disruptive infrastructure in their own communities without working to make sure that everyone has a say in how things are done.

You can remember before the shift in energy ownership when big oil, gas and other energy companies had held huge amounts of political power. They had done everything they could to water down IPCC and country-specific climate and environmental plans (4). They'd pushed for huge amounts of government money to go into technologies that would allow them to keep making massive profits at the expense of the climate, and the sustainability of life on earth. They'd destroyed livelihoods by contaminating natural environments, and interfered in political systems that weren't amenable to their interests. They were remarkably successful in lobbying politicians to get what they wanted. They were also pretty good at influencing the general public through social and traditional media to align ideas like individual car ownership with values like freedom. It had been insidious and manipulative, but it had worked.

As people had started to take control of energy systems, powerful energy interests had fought back. In the US, the Koch Brothers, a powerful family built on oil wealth, had lobbied state governments to try to block the solar energy revolution (5). They were initially successful in some places but it had turned out to be impossible to hold back the tide. As the costs of solar and wind fell, people realised how easy it was to participate in the energy system. Broad participation started to redistribute the economic benefits from energy production away from corporate shareholders and CEOs and to local communities. After that, there was no turning back.

The explosion of distributed energy ownership ate away at the vast political power of former energy interests. That had begun even before the pandemic started (Brisbois, 2020). Once those companies no longer controlled the ability to turn on the lights, or to get to and from work, it was like they lost their teeth. You can remember the slow shift of the government away from public climate and energy policies that felt like they were solely focused on profits for big companies. It had required a huge amount of public pressure, including mass movements in the street to see the initial changes (Temper et al., 2020). However, the loss of market share to non-profit oriented interests meant that governments suddenly had much more freedom to actually govern in the interests of the people.

Some of the energy system was still centralised. In every country, the mix was a little bit different depending on the resources they had available, and the local context. In the end, it turns out that people do like the security of being connected to the grid. It also just makes sense to use clean centralised sources like hydro and offshore wind farms where there are abundant resources, and where development doesn't destroy local livelihoods. The mix means that there's enough energy to power even big cities where demand is high, but the space available for generation is limited.

Local microgrids that can be disconnected from the main grid have also made energy systems much more stable. This is important because we'd already locked in global temperature increases before taking actions to stabilise emissions. There are more frequent and intense storms than there ever used to be. However, a decentralised and compartmentalised grid means that the whole system never goes down. It also means that things get back up and going much more quickly and people are able to share their generation capacities.

You could go on, but you're anxious to get out on your bike and get to work. It's a beautiful day out there...

Creating our future

In any future scenario, including the options presented above, there are trade-offs. Our current way of life also requires trade-offs – we've just gotten used to them. For example, it's currently normal to trade-off our mental and physical health for a private automobile-based commute. We trade off control over our governments for access to cheap, disposable products. We trade off a liveable planet for the comfort of the status quo. The key in picking any option, and also in examining our current situation, is to explicitly identify who wins and who loses, and what we gain and what we must give up, in each scenario.

It's also helpful to remember that fear is a powerful motivator. A lot of conversations about the ongoing energy transition are rooted in fear. The catastrophic consequences if we don't change our energy system is one major source of fear. The fear of disruptive change to our lifestyle is another. Research on the psychology of emotions has established that fear often makes us retreat into the values and worldview that we already hold (Pyszczynski et al., 2020). This means that, those inclined to climate action are likely to be more motivated in the face of these fears. However, the part of society that is reluctant to accept the realities of climate change are likely to retreat into denial and sometimes even more energy consumptive activities. We can see these dynamics playing out around the world.

This means that successfully transitioning our energy system can't just be a project of fear. We need other stories. We need to imagine a world where life doesn't just use less energy, but where it is objectively *better*. We need people to sign up to an energy transition not just because they have to, but because they want to. Luckily, there are lots of ways a different way of life will be better.

The overlap between energy, working and social lives, air quality, the natural world, mental and physical health, systems of production and consumption, and the quality of our political systems means that there are a lot of vectors along which we can make things better. There are a thousand different tentacles to this transformation and we probably need to work across all of them to reach the end point. And while this seems overwhelming, every single one of those vectors is also an entry point to helping transform our energy system into something that will work better for everyone.

References

Aronoff, K., Battistoni, A., Cohen, D. A., & Riofrancos, T. (2019). *A planet to win: why we need a Green New Deal*. Verso Books.

Baldwin, E. (2020). Why and how does participatory governance affect policy outcomes? Theory and evidence from the electric sector. *Journal of Public Administration Research and Theory, 30(3),* p. 365-382.

Brisbois, M. C. (2020). Shifting political power in an era of electricity decentralization: Rescaling, reorganization and battles for influence. *Environmental Innovation and Societal Transitions, 36,* p. 49-69.

Geels, F.W. (2004). From sectoral systems of innovation to socio-technical systems: Insights about dynamics and change from sociology and institutional theory. *Research Policy, 33(6-7),* p. 897-920.

Hook, A., Sovacool, B.K., & Sorrell, S. (2020). A systematic review of the energy and climate impacts of teleworking. *Environmental Research Letters, 15(9),* p. 093003.

Mytton, O. T., Panter, J. & Ogilvie, D. (2016). Longitudinal associations of active commuting with wellbeing and sickness absence. *Preventive medicine, 84,* p. 19-26.

Pyszczynski, T., Lockett, M., Greenberg, J., & Solomon, S. (2020). Terror Management Theory and the COVID-19 Pandemic. *Journal of Humanistic Psychology,* 0022167820959488.

Salvi, B. L., & Jindal, S. (2019). Recent developments and challenges ahead in carbon capture and sequestration technologies. *SN Applied Sciences, 1(8),* 885.

Szulecki, K. (2018). Conceptualizing energy democracy. *Environmental Politics, 27(1),* p. 21-41.

Temper, L., Avila, S., Del Bene, D., Gobby, J., Kosoy, N., Le Billon,

P., & Walter, M. (2020). Movements shaping climate futures: A systematic mapping of protests against fossil fuel and low-carbon energy projects. *Environmental Research Letters, 15(12),* 123004.

Thilmany, D., Canales, E. Low; S.A., & Boys, K. (2020). Local Food Supply Chain Dynamics and Resilience during COVID-19. *Applied Economic Perspectives and Policy.*

Timpe, C. (2009). Smart Domestic Appliances Supporting the System of Integration of Renewable Energy. *Report of the Intelligent Energy Europe "Smart Domestic Appliances in Sustainable Energy Systems (Smart-A)" project.* https://ec.europa.eu/energy/intelligent/projects/sites/iee-projects/files/projects/documents/e-track_ii_final_brochure.pdf

Online resources

1. https://www.theguardian.com/us-news/2020/jul/27/fossil-fuels-oil-gas-industry-police-foundations
2. https://www.climateinteractive.org/ci-topics/multisolving/
3. https://theconversation.com/work-less-to-save-the-planet-how-to-make-sure-a-four-day-week-actually-cuts-emissions-124326
4. InfluenceMap, Big Oil's Real Agenda on Climate Change, March 2019, https://influencemap.org,/report/How-Big-Oil-Continues-to-Oppose-the-Paris-Agreement-38212275 98aa21196dae3b76220bddc
5. https://www.rollingstone.com/politics/politics-news/the-koch-brothers-dirty-war-on-solar-power-193325/

Chapter 11:

Moving Beyond Climate Change Denial and the Anthropocene's Ecocidal Logic

Martin Hultman

Originally published as: Hultman M (2020) Politics at the End of the Anthropocene. *Georgetown Journal of International Affairs,* 20 April.

Editor's note: The article that we republish here, under a new title, was originally posted early during the pandemic, yet it remains, I feel, one of the best pieces of thinking and writing to appear during the Covid 'era'. I am thus delighted to be able to include it here, in a collection into which it fits so well.

"For over thirty years the scientific knowledge of global climate change has been on the political and public agenda, but today, we are further from dealing with the root causes than ever before. Emissions from burning fossil fuels and other sources of greenhouse gases are on the rise, while extractive industries and far-right wing political parties engage in climate change denial."

The above introduction was supposed to be the opening of a text focusing on the overlap of the climate crisis, right-wing authoritarianism, and the origins of both in our fossil fuelled global economy. I was supposed to bring together all of my interdisciplinary knowledge on these subjects from the research we are carrying out at the Center for Studies of Climate Change Denial in Sweden. Our findings have just been published in a book with the title Heated – Democracy in the Period of Climate Crisis. I had for a few weeks written down my judgments of what to do if we are to find safe ground when lowering the greenhouse emissions. But nobody foresaw the outbreak of Covid-19.

The whole world today is in a state of flux, and will be for months, if not years, to come. At the same time, many of the measures taken to prevent the spread of Covid-19 (lockdowns, closed borders, etc.) are in line with the authoritarian shift taking place in some of the largest countries in the world. While it feels like a new reality, this is not a unique time in human history. If we are to handle this triple crisis of climate change, authoritarianism, and diseases in a way that actually shifts us on a path towards healing Earth and its Earthlings—creating a center of gravity attractive enough to shift the field of energy towards a sustainable future—we need to recognise the historical patterns that have led us to this point. When researchers just have named our current geological period as the Anthropocene, maybe our best thinking should be about how to end this idea of human exceptionalism at the core of such a term?

Ecocidal logic of late-Anthropocene

A bit more than a decade ago, four global societal processes came together. The first three trends—high awareness of the climate crisis, economic recession, and a pandemic (H1N1)— resulted in the fourth outcome, as less coal, oil, and gas was extracted and burned, leading to decreasing carbon emissions. At the time, the political response was to "fossil fuel up" the economy (e.g; shale gas in the US, tar sands in Canada). Today, another reaction is needed.

The fossil fuel industry cannot "save us" this time since it has for decades been a merchant of doubts spreading contrarian climate science while it has pumped and mined oil, gas, and coal. They have successfully created an "ideological" climate change denial among right-wing authoritarian nationalist political parties around the globe (led by Trump, Bolsonaro, and Morrison, to name a few). When the youth climate justice movement brought climate into politics, the "climate denial machine" (as termed by professor Riley Dunlap) again went into a higher gear. This engine is well funded, well established, and connected to right-wing nationalist political agendas. Its power is widespread. The maintainers of an audience for the messages sent out by the climate denial machine is quite a homogenous group of older men with conservative values. Challenged by the straightforwardness of young people in the climate movement in 2018 and 2019, men with conservative values who were part of right-wing nationalist parties enacting "industrial/breadwinner masculinities" have reacted with anger, fear, and confrontation. This reaction is connected to the structures of industrial modernisation and our fossil fuel based global economy which has demanded these types of values and practices by especially Western men since the 17th century.

We are now in the middle of a perfect storm which was (as in 2009) proceeded by widely shared knowledge regarding the climate crisis. Added to this is a pandemic (as in 2009). In this recession, less coal, oil, and gas is extracted and burned, leading to decreasing carbon emissions; however, the global economy might be "fossil fuelled up" again if transformative policies are not put in place.

Where to now?

If we are to leave anthropocentric extractive logic behind, laws and norms need to be changed—as norms and laws were changed when we entered the Anthropocene. Just as the industrial revolution was made possible not only by technological innovations, but also very much by shifts in values and laws making extractivism the new normal, our global climate emergency needs similar shifts towards glocal care for our planet. Two suggestions to that end are outlined below.

First is the need to try out laws that protect the planet. For many years, scholars from a diverse set of fields such as law, sociology, Indigenous studies, and gender studies have laid out the vision and practice of Rights of Nature, arguing that "nature" in and of itself should be part of political decision making. These ideas are more acute and important today than ever, not least because it seems that only including "nature" as a resource in policy making or as a limit not to be crossed is failing us as a species. Rights of Nature are today inscribed in the constitutions of Ecuador and Bolivia; additionally, rivers in New Zealand and Lake Erie in the United States have been granted rights. This approach could also be implemented globally through the Rome Statute. The Rome Statue underpins international law and, combined with a fifth statue in the form of an End Ecocide Law—similar to the law against Genocide—could bring court cases to the International Criminal Court in the Hague. For example, a case against those responsible for the Great Barrier Reef bleaching may be possible.

Second is the need to change the gender norms that shape men into "industrial/breadwinner masculinities." This way of framing maleness is today present mostly in the same aforementioned cohort of climate change deniers and is failing both men themselves and the broader societies they live in; just look how Scott Morrison handled the fires in Australia, or how Jair Bolsonaro acted with Covid-19. A shift is needed toward masculinities with greater care for men themselves, as well as for women, youths, societies, and the Earth. In scholarship and education, younger and more aware men are turning toward what has been termed "ecological masculinities" as a way to be just and careful with all humans and non-humans alike. Inspired by academic rigour from the traditions of ecological feminism and feminist care theory, "ecological masculinities" enacting caring encounters with self and others, recognises our material interconnectedness with humans and other-than-humans alike, identifying the costs of male domination as well in pro-feminist solidarity creating a just society for all bringing us back to Earth.

There is a need in this moment to be innovative and work with solutions that can make a big difference. Rather than "fossil fuel up" the climate change denial machine once again, new forms of laws and transformations of gender norms are two solutions that will create a new path for humanity.

Chapter 12:

Better Environmental Communication in a Post-Covid World

Dr Sibylle Frey, with contributions from
Max Winpenny and Brittany Ganguly

The rise of populism around the world is a challenge for action on climate change and other environmental crises. Given a probable climate-induced social collapse during this century - perhaps with Covid-19 as a precursor - it is even more important to understand why these forces deny climate change and to bridge the gap between science and collective environmental action. We need different methods of communication, based on shared values that will support a radical change in our thoughts, beliefs, and ultimately, of our outdated social and economic system. Touching on examples from several countries and lessons from neuroscience, we explore communication's role in the environmental dilemma and how it could be improved to inspire positive action.

What are we up against?

Mounting right-wing populism in the Americas, Asia, and Europe is a backlash to multilateral action on climate change and other environmental problems. Understanding how these groups perceive climate change and influence environmental action is crucial for effective communication and the design of inclusive policies.

In the United States, for example, the predominant forces behind climate change denial are big money, vested interests, and the evangelical right. The organisational, strategic, and financial power of America's rich controls much of the news media. Fossil fuel and other corporations have been instrumental in undermining climate science and negotiations, sowing seeds of doubt where scientific consensus had been reached. Having established their own think-tanks, in particular the secretive Council for National Policy – elsewhere described as a 'pluto-theocracy' – has been pivotal for the American right for mastering politics and the media and its influence is global.

So far, within Europe's populist right-wing parties, climate change-scepticism is not as ingrained in dogma and identity as in the U.S. and has less of the strategic denial to preserve the status quo. A recent study on Europe's right-wing populist parties finds that these have three key features in common: first, most of them oppose climate policies, however, grievances are mainly rooted in economic or social injustice; second, climate action is perceived as a globalist issue and a liberal-elitist concept underpinned by mistrust towards multilateralism and international institutions and third, despite this, these parties tend to support local environmental initiatives. There is some sort of "green patriotism" which backs nature conservation but not climate action. Given the growing injustice in a fast-changing, globalised world, and its many flopped climate policies the faulty communication strategies of climate policies have come under scrutiny. So, what are the solutions?

Designing better environmental communication strategies and policies

Goodbye old power models – hello local concerns and shared values

In the US in particular, environmental communication must provide a counterweight and counter-spin against the skilled and united front of the populist/evangelical right and get equally organised and smart. The entrenched systems of strategic denial have to be exposed and this will require immense effort and investment. However, although the US news media is owned by 15 billionaires not all are right-wing. For example, Michael Bloomberg donated more than $5 billion to gun control and climate change and funded the Democrats in the 2020 presidential election. The main issue, however, is to break up the manipulative power model that governs the news and social media. Some believe that this process has already started.

The lessons from Europe show that policymakers can no longer treat climate action solely as a technical fix. Climate policy affects people and must therefore be embedded in credible and fair social policy and connect to local concerns. Inevitably, deep social transformations create conflict; there will be winners and losers and this cannot be glossed over. A common theme among Europe's populist parties is their rejection of pluralism – claiming that they are the only parties representing those who feel betrayed and neglected by the "corrupt and demoralized elites" that steer the ship. This elite is therefore also responsible for the mounting mistrust in science. Populism, climate scepticism, and hostility may all sound bad but also shows that communication must be mutual, aware of people's values and needs, and embedded in the local context. Whether we like it or not, we have to work with the "dark side" and acknowledge that there may be specks of truth in some of the populist's talk.

We have to listen and take people's concerns seriously, discuss drawbacks, trade-offs, and uncertainties honestly and transparently, and deflate fake news without reinforcing them. We need success stories that implant trust in political change. Since populist parties typically frame their programmes around independence, economic development, homeland, fairness, and nature, they offer opportunities for finding mutual ground. We may not reach those who do not want to be reached (like those in strategic denial) but we can change the narrative for climate change to bridge the gap between the privileged and those who feel left behind.

Ethical implications

Storytelling is a powerful tool that needs to be used responsibly. Controlling the narrative is strategically important but so is questioning whether it represents a broader truth or a narrow purpose. Environmental communication must tell things as they are without "beautifying" but also without blaming and shaming. Highlighting the damaging functions of institutions and the key individuals who serve them is crucial but so is recognising that we have a destructive economic system that does not serve anyone in the long run.

Given a probable climate-induced societal collapse in the near future and the fact that scientists and institutions usually err on the side of least drama, not everyone agrees to what extent environmental impact scenarios should be debated in public. Some fear that doomsday narratives are irresponsible and may disillusion and disempower people, thus underplaying the adaptive capacity of societies to cope. Others argue that we should not censor people's ability of sensemaking because doing so would restrict them from reaching their own conclusions. Rather, we should consider the implications of a climate-induced social collapse in all their depth (such as mass migration, disease, war, and starvation) to connect emotionally to what is happening; this will help to prepare ourselves. What effect will the communication of such a catastrophe have on people's psychology? Will it result in despair or trigger strengthened action for remediation? Right now, there are no conclusive answers; however, from ancient traditions, terminally ill patients and people who have experienced immense loss (like the American Indians, for example) we know that the feeling of hopelessness is an important first step towards a new understanding of self and the world, and one which can open up space for alternative hopes.

Lessons from Covid-19

Along came a virus. Covid-19 affects most of us and its wider repercussions will do so for some time. The virus has shown how fragile our civilisation is and reminded us of our mortality. However, Covid-19 also triggered unprecedented levels of collective action by governments, communities, and citizens. This collective action, however, also depends on whether the communication of public health strategies is effective - with politicians' credibility ratings rising or falling accordingly.

Rather than frantically rushing back to a pre-Covid era (if that is at all possible) we should heed the pandemic as a warning shot. The virus removed our blinkers and exposed that the crisis is of our own making: it is us who have created the ideal environment for disease to evolve and spread. This includes climate change, habitat destruction, the way we treat and farm animals, and an overpopulated planet. Covid-19 is a primer on how to prepare for future shocks. Grim perhaps, but this is our chance to prepare for what we urgently need - a new agenda for rebuilding our societies and economies, and for restoring ecosystems and agriculture. To accomplish this, we need imaginative, strong narratives built on credible and ethical foundations and which help people re-connect with nature and each other.

The (neuro)science behind communication: emotions are the fuel of behaviour

"Thou shalt not alienate"

Politicians and scientists are notorious for using technocratic language when they talk about climate change. This neglects social realities and alienates people. Facts and figures all have their place but it's "stories, not bar charts, that change the world." MRI scans have shown that storytelling or watching a film engages different parts of the brain, thereby affecting our intellect and emotions. When listeners hear a good story, their brain activity mirrors that of the speaker. These mirror-neurons enable us to feel empathy - the ability of "being in the shoes of the other." The listener's brain forms a neural coupling with that of the speaker and is pulled into the moment - both brains just "click." The greatest level of neural synchronisation and understanding occurs when the listener's brain can guess what might happen next. The listener can identify with the motivations of the storyteller and take a perspective. Such an engagement of hearts and minds allows people to shift attitudes: when they come out of the narrative they are more likely to think about an issue differently.

Empathic confrontation

The multi-dimensional phenomenon of empathy is widely used in psychotherapy, marketing, business management, and film-making. Principles of Empathic Confrontation (EC) originate in conflict mitigation and variations thereof are applied to many settings. EC is assertive rather than passive or aggressive, holding those accountable while not attacking them and avoiding shame and opposition. For example, rewilding efforts have shown greater success where local people were listened to and engaged early on in the process. EC rests on

1. Building a connection with others,
2. Labelling wrong behaviour,
3. Acknowledging why the behaviours developed,
4. Planning for collaborative correction.

EC intends to move the recipient(s) from the red zone of being unreceptive to feedback (for example due to feeling shamed) to a more receptive stage where they are open to feedback and behaviour change (green zone; figure 1).

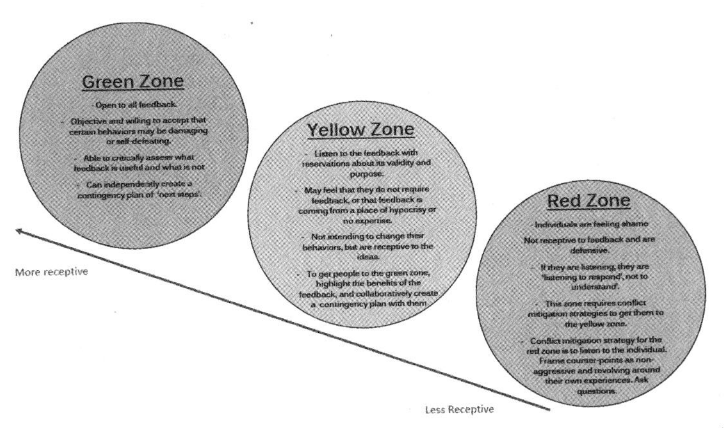

Figure 1. Zones of receptivity to feedback (image courtesy of Max Winpenny).

For example:

1. Building a connection: *"These times are challenging for all of **us**"* (opening statement to connect; the speaker takes responsibility which makes it easier for the listener to take responsibility).

2. Labelling: *"It is right to look at those in charge and say that they aren't doing enough. We can hold them accountable for that. But first, **we** need to hold ourselves accountable for our own actions."* (The use of "us," in number 1, and "we" creates joint accountability and understanding that this is difficult and not anyone's 'fault.' It creates connection between listeners and the speaker, and develops a sense of collaborative hope rather than individualistic competitiveness.)

3. Acknowledge/Empathise: *"This won't always be easy… but the most we can ask from each of us is to reflect on how we act and try to do better."* (Acknowledgement of why the behaviour occurred.)

4. Collaborative correction: *"This is going to take group action… With your ideas, your input and your effort, there is very much a way towards a better future."* Or, in Barack Obama's words, **"yes we can."** (Actionable when working towards a common goal, powerful because 'yes' is positive, 'we' is empathetic, and 'can' is actionable.)

Examples of effective storytelling

Hearing a real-life story about someone we can identify with, for example, a farmer who has lost his entire crop during the last heatwave or a patient who survived Covid-19, can help people to weigh up the consequences and is more likely to gather support for action than just presenting the facts. Instilling positive emotions through stories of change is key to motivation, empowerment, and engagement. Here are some other examples:

POPULATION GROWTH:

Facts and figures:

"The world's population is projected to grow from 7.7 billion in 2019 to 8.5 billion in 2030 (10% increase), and further to 9.7 billion in 2050 (26%) and to 10.9 billion in 2100 (42%)" - United Nations World Population Prospects 2019.

Engaging:

"Condoms are weapons of mass protection"

"[...] and their kids were doing it in schools too - we had air races with condoms, we had children condom-blowing championships, and before long the condom was known as the girl's best friend. In Thailand, for poor people, diamonds just don't make it. And what happened? From seven children to 1.5 children between 1974 and 2000 [...]. And that's the case of everyone joining in - we didn't have a strong government, we didn't have lots of doctors but it's been everyone's job to change attitude and behaviour." - Mechai Viravaidya, former Thai minister and chairman of the Population and Community Development Association (PDA) who led the successful campaign to reduce population growth, poverty, and later HIV infections in Thailand.

SIXTH MASS EXTINCTION:

Facts and figures:

"For terrestrial and freshwater ecosystems, land-use change has had the largest relative negative impact on nature since 1970, followed by the direct exploitation, in particular overexploitation, of animals, plants and other organisms, mainly via harvesting, logging, hunting and fishing"- IPBES Global Assessment Report 2019.

Engaging:

"When people think of extinction it is an imaginary tale mostly told by conservationists. But I have lived it - I know what it is. I am caretaker of the Northern White Rhinos. We only have two left on this planet. They are mother and daughter. This is Najine, the mother. She is 30 years old. She's very quiet. And her daughter is Fatu: she's 18 years old. She's pretty much like a young teenager; she's a bit unpredictable and can be feisty sometimes, especially if she wants something [...] I've seen these beautiful rhinos count from seven down to two. They're here because we betrayed them. And I think they feel it, this threatening tide of extinction that's pushing on them. When Najine passes away she'll leave her daughter Fatu alone forever. The last Northern White Rhino. And their plight awaits a million other species." - James Mwenda, conservationist at Ol Pejeta Conservancy, Kenya, in: Extinction: the facts - BBC September 2020.

"Over the course of my life, I've encountered some of the world's most remarkable species of animals. Only now do I realize just how lucky I've been. Many of these wonders seem set to disappear forever. We're facing a crisis and one that has consequences for us all. It threatens our ability to feed ourselves, to control our climate, it even puts us at greater risk for endemic diseases such as Covid-19. It's never been more important for us to understand the effects of biodiversity loss. Of how it is that we ourselves are responsible for it. Only if we do that will we have any hope of averting disaster [...] What happens next is up to every one of us." - David Attenborough, in: Extinction: the facts; BBC September 2020.

CLIMATE CHANGE

Facts and figures:

"We will need a mix of adaptation and mitigation measures to meet the challenge of climate change." - World Meteorological Association.

Engaging:

"Right here, right now is where we draw the line. The world is waking up. And change is coming whether you like it or not [...] Together and united, we are unstoppable." - Greta Thunberg.

Summary and conclusions

Effective environmental communication is essential for engaging people, shaping individual and public opinion, ensuring accurate information is shared widely, and achieving societal transformation.

Understanding the causes of science scepticism and climate change denial is vital for developing effective communication strategies and for achieving societal change towards a sustainable world. Such strategies have to be built on shared values, identity and inclusion to avoid disillusionment – a key factor in the surge of populism.

The ethical implications of environmental communication include the extent to which a climate-induced social collapse should be discussed in public. Perhaps, if we deeply immerse ourselves in the collapse and connect emotionally with what is happening around us we can prepare ourselves better for the things to come.

Against this backdrop of collapse, Covid-19 fired a warning shot to focus our attention on what we urgently need: a new agenda for rebuilding our societies and economies, and for restoring our ecosystems. Especially for a post-Covid world, we have to find new ways to deal with reactionary forces. This requires communication strategies built on sound foundations and with narratives that trigger the imagination, empower, and engage all parts of society.

Neuroscience proves that such narratives can change people's hearts and minds. The reason why stories can be so powerful is that they synchronise our brains at the neural level, triggering empathy and fuelling behaviour. Hearing real-life stories about someone we can identify with is therefore more likely to gather support for action than presenting the facts alone. Or, in Jane Goodall's words, "evidence is not enough - you have to go to the heart."

At the core is this: if we truly want to move others the most important point is that we need to deal with the outside world coming from a place within. Only then can we create a world where humanity is in sync with the rest of nature. Yes we can!

References and further reading

1. Millennium Alliance for Humanity & the Biosphere, Stanford University, San Francisco, CA, USA. https://mahb.stanford.edu/

2. Definition of right-wing populism: *"Right-wing populism [...] is a political ideology which combines right-wing politics and populist rhetoric and themes. The rhetoric often consists of anti-elitist sentiments, opposition to the perceived Establishment, and speaking to the "common people"* (2020). *Wikipedia.* https://en.wikipedia.org/wiki/Right-wing_populism (Accessed September 2020). For easier reading, in this chapter we mainly use the term populism.

3. Schaller, S. & Carius, A. (2019). *Convenient Truths: Mapping climate agendas of right-wing populist parties in Europe.* Berlin: adelphi.

4. These 15 Billionaires Own America's News Media Companies (2016). *Forbes Media LLC.* https://www.forbes.com/sites/katevinton/2016/06/01/these-15-billionaires-own-americas-news-media-companies/ (Accessed October 2020).

5. Lawrence, F., Pegg, D., Evans, R. (2019). How vested interests tried to turn the world against climate science. *The Guardian.* https://www.theguardian.com/environment/2019/oct/10/vested-interests-public-against-climate-science-fossil-fuel-lobby (Accessed November 2020).

6. Council for National Policy. (2020). *Wikipedia.* https://en.wikipedia.org/wiki/Council_for_National_Policy#Conferences_

and_political_plans (Accessed October 2020).

7. Nelson, A. (2019). *Shadow Network.* New York, NY: Bloomsbury.

8. The Politics of the Solar Age – A MAHB Dialogue with Author and Global Futurist Hazel Henderson. (2020). *MAHB Blog.* https://mahb.stanford.edu/blog/the-politics-of-the-solar-age-a-mahb-dialogue-with-author-and-global-futurist-hazel-henderson/ (Accessed November 2020).

9. Müller, J.W. (2017). *What is populism?* London: Penguin Books.

10. Bray, R. (2002). *SPIN works! A media guidebook for communicating values and shaping opinion.* San Francisco, CA: Independent Media Institute.

11. *What happens in the brain when we hear stories?* Uri Hasson at TED (2016). https://blog.ted.com/what-happens-in-the-brain-when-we-hear-stories-uri-hasson-at-ted2016/ (Accessed October 2020).

12. Bendell, J. (2018). Revised 2nd Edition Released July 27th 2020. Deep adaptation: a map for navigating climate tragedy. Institute for Leadership and Sustainability (IFLAS) Occasional Papers Volume 2. University of Cumbria, Ambleside, UK. (*Unpublished*).

13. Brysse, K., Oresekes, N., O'Reilly, J., et.al. (2013). Climate change prediction: Erring on the side of least drama? *Global Environmental Change 23, p:* 1.

14. Why Hope Is Dangerous When It Comes to Climate Change. (2020). *Slate.com.* https://slate.com/technology/2017/07/why-climate-change-discussions-need-apocalyptic-thinking.html (Accessed August 2020).

15. Daily, G.C. & Ehrlich, P.R. (1996). Global Change and Human Susceptibility to Disease. *Annu. Rev. Energy Environ. 21,* p. 125–44

16. Praszkier, R. (2016). Empathy, mirror neurons and SYNC. *Mind Soc 15,* p. 1–25.

17. Defend Your Research: I Can Make Your Brain Look Like Mine. (2010). *Harvard Business Review.* https://hbr.org/2010/12/defend-your-research-i-can-make-your-brain-look-like-mine (Accessed October 2020).

18. How Stories Connect And Persuade Us: Unleashing The Brain Power Of Narrative. (2020). *NPR The Coronavirus Crisis.* https://www.npr.org/sections/health-shots/2020/04/11/815573198/how-stories-

connect-and-persuade-us-unleashing-the-brain-power-of-narrative (Accessed October 2020).

19. Your Brain On Storytelling. (2020). *NPR Shortwave*. https://www.npr.org/2020/01/13/795977814/your-brain-on-storytelling (Accessed November 2020).

20. Roeser, S. (2012). Risk communication, public engagement, and climate change: a role for emotions. Risk analysis: an official publication of the *Society for Risk Analysis* 32, p. 6

21. Stephens, G.J., Silbert, L.J., Hasson, U. (2010). Speaker–listener neural coupling underlies successful communication. *PNAS 107* p. :32

22. Farrell, J.M. & Shaw, I.A. Empathic Confrontation in Group Schema Therapy. In: Reiss, Vogel, eds. *Empathic Confrontation (in German)*. Frankfurt: Beltz.

23. Sutton, A.E. (2015). Leadership and management influences the outcome of wildlife reintroduction programs: findings from the Sea Eagle Recovery Project. *PeerJ*, 3, e1012.

24. U.S. Department of the Interior Fish and Wildlife Service. (1993). *The reintroduction of gray wolves to Yellowstone National Park and central Idaho*. Draft environmental impact statement.

Mother Earth needs us to keep our covenant. We will do this in courts, we will do this on our radio station, and we will commit to our descendants to work hard to protect this land and water for them. Whether you have feet, wings, fins, or roots, we are all in it together.

Winona LaDuke
Environmental activist

Chapter 13:

Afterword
On Rational and Sacred Ground: Envisioning a Better Future for All Life

Eileen Crist

Originally published as: Crist E (2020) On rational and sacred ground. *Earth Tongues*, 31 August. Available at: https://blog.ecologicalcitizen.net/2020/08/31/on-rational-and-sacred-ground/ (Accessed February 2021).

Editor's note: In pulling together this collection, I was determined to include a piece by Eileen Crist, who is an empassioned and important ecological writer. The piece that we publish here, taken from an exiciting new multi-contributor blog called *Earth Tongues*, felt like a most fitting afterword for this collection. As Eileen concludes here, the time has come for humanity to "secede from the dominant no-limitations, life-destroying civilisation, and build alternative communities of human life on rational and sacred ground."

We are in the midst of the Sixth Extinction event, which, if completed, would entail losing 50-75% of all species within the century. Yet most media and politicians observe a resounding silence about this imminent, unthinkable catastrophe.

Which begs a question. Do people actually believe that humanity can cause a mass extinction and there will be no consequences? I am not talking only about consequences for human physical well-being, which will be copious. I am talking about an everlasting legacy of sorrow to human posterity. The sorrow of robust, beautiful life forms exterminated forever, pointlessly and under our watch. The sorrow of the human identity of planetary colonizer having triumphed over the human identity of loving participant. These are soul consequences. If the word "soul" troubles you, then please simply follow your intuition into a future of a completed mass extinction, and allow yourself to *feel* that reality.

We must do everything we can to stop the Sixth Mass Extinction.

To that end, we must take immediate action to downscale the human enterprise and pull back from the natural world. Just powering a burgeoning global civilisation with renewable energy will not save the diversity, abundance, and complexity of life on Earth. To save life's splendour, we must additionally and more importantly address the "burgeoning" aspect of human affairs.

oOo

Earth's integrity is degrading rapidly and at multiple levels with each passing day. At the same time, the liable patterns of human consumption and of population dynamics around the globe are grossly uneven, making generalisations about who is responsible, and how to proceed as an international community, extremely challenging.

Even so, certain generalisations are possible about how to move forward that we can agree on. Let me unfold that argument.

Most of the world's poor are seeking to join the wealthy in terms of income expendability and comfortable consumption. The Earth, however, cannot offer food, water, energy, and materials—for a projected 9 to 11 billion people—at an American standard of living. A world of 10 billion "Americans" would entail either the collapse of global civilisation (if unsuccessful), or human tyranny over the face of the Earth (if successful). I am not sure which outcome would be worse.

Thus, an American standard of living is not viable, except (as presently) under conditions of acute human inequality. Since inequality is unacceptable, we have arrived at a first conclusion we can agree on: an American standard of living is unacceptable. We need to roughly equalise consumption among all people, at a *lower* than American level, while also preserving a biodiverse planet.

What might that standard of living be?

Let's again start with what cannot be rejected. *Electrification.* Humanity has opted in and will not (voluntarily) opt out. Electricity is here to stay, even as the global community must learn to conserve it and make it with cleaner energy.

Right off the bat, having electricity entails consumption beyond the bare basics. Electricity demands a materials-intensive energy infrastructure and it enables, via the internet, swift trade flows and instant purchasing power. Additionally, electricity is the prerequisite for enormous amounts of plugged-in "mundane" stuff, like refrigerators, phones, washing machines, TVs, PCs, e-bikes, electric cars, *and so on*, which are based on extraction, use energy to make and operate, and produce prodigious waste at both the production and consumption ends.

Overall, electricity supports unprecedented movement of money, information, advertising, transactions, and material flows. All these flows come with immense consequences for the Earth. *That's why an electrified civilisation has to be a low population civilisation.* But I'm running ahead of myself.

An electrified human life, which almost everyone wants, is well beyond the basics. But we might also think about the matter of consumption from another angle, namely, what people tend to desire. What most people desire is some level of comfortable consumption.

Indeed, we can divide goods and services into three categories: basics, comforts, and luxuries. While it is by no means simple to delineate them crisply, we can provisionally agree that all people, for starters, should have the basics and that luxuries (for the sake of the Earth and human equity) have to go. We can also concur that when it comes to goods and services, most humans gravitate toward the category of "comforts."

To recap. Thus far we have agreed about the imperative of an equitable standard of living that is somewhere in the comfort category, one, because it's electrified, and two, because it's what people want.

Having converged on the above—if only because reality compels us—we can turn to the big variables that would determine *desirable* global consumption. Desirable global consumption can be defined as one that supports a thriving biodiverse planet within which humanity enjoys an electrified material culture. The determining variables of such consumption are two: the total number of equitably consuming humans and the design of economic life at global and local levels.

Let's start with the economy, drawing on some degrowth thinking: what kind of economic life could offer humans an electrified way of life on a biodiverse planet? The motto "Reduce-Reuse-Recycle" popularised in the 1970s still goes a long way. In production, "reduce" means lowering throughput. This is achievable by increasing efficiency (using less to make the same) and by absolute reduction (using less to make less). Both are necessary. We need greater efficiency and we need what ecological economists call a less busy economy. A less busy economy means that we work less—enjoy a shorter workweek—in order to make less stuff, use less energy, and produce less waste.

Not only will this economic design be far more ecologically benevolent, it will free time to devote to what is truly fulfilling—time for nature, family, community, hobbies, and self-realisation.

The production side also must use recycled materials. Glass, metals, wood, high-tech parts—whatever can be recycled should be recycled. What cannot be recycled must be substituted (especially if it is dangerous). Synthetic pesticides, fertilizers, and plastic are excellent examples of stuff that should not be made in the first place.

Turning to the consumption end of the economy, what people consume will of course be tempered by the design of production. Additionally, we must consume more locally, value more handmade products, reuse/fix/share/pass down, and eat lower on the food chain. In other words, we need to create a new global *culture* of consumption.

So far in the quest to downscale the human enterprise, we've arrived at the need for an equitable (electrified) standard of living, organised around a less busy economy (that is more local and as circular as possible), in a world where eating far fewer animal products implies far fewer livestock.

That brings us to the last big variable shaping aggregate consumption and waste under the above scenario, namely, how many people there are on planet Earth. Let's start at an intuitive level: under the just-stated conditions, would there not be yawning difference between, say, 3 billion people and 9 billion people in terms of ecological impact?

But "intuitive" is not good enough. We need to sharpen our reasoning by foregrounding something that so far has remained present but backgrounded: the more-than-human world, and preserving its magnificent diversity of species, subspecies, populations, genes, ecologies, migrations, and biomes.

If we want to stop the Sixth Mass Extinction, help absorb much of the life-wrecking carbon we've unleashed, and (beyond these pressing goals) support the natural world to flourish again, then we need to protect it, restore it, and give it space. Large-scale, connected space.

Perhaps we might agree, as some initiatives are proposing, that 50% of the land and 50% of the ocean is fair to give back to free nature. *Fair*, given that one species gets half and the other 5 to 10 million species get half. (To clarify, "half" does not mean giving nature rock, ice, and marginal lands, but at least half of all representative ecosystems.) That recalibration of Earth's geography places definite limitations on infrastructural sprawl and even demands undoing many existing infrastructures.

But what truly sharpens the parameter of human population size—in such a world of generously freed nature—is *food production*. A turn away from industrial food production to organic, diversified agro-ecology is the only means to heal the ecological devastation of industrial food. Such a turn implies abolishing synthetic fertilizers. Synthetic fertilizers, however, have bankrolled the human population explosion. (Over 40% of us would not be here without them.) Therefore, ending industrial food implies gradually achieving a lower population.

Not only should lowering our numbers be done within a human-rights framework, but expediting certain human rights for all people is the precise pathway to do so. These universal rights are: family-planning services for all, accessible and affordable; all girls and young women in school at least through secondary education; and connected with the previous, zero tolerance from the international community of "child brides."

In sum, scaling down and pulling back entail: creating an interconnected electrified civilisation; designing slow, local, and mostly circular economies, at an equitable (say roughly European) standard of living; and embracing and enabling a generously freed natural world. Between human habitats and wild habitats, the "middle landscape" where we make our food must be de-industrialised and remade as part of the biosphere—friendly to life, mimicking life, enhancing life, loving life.

Those of us who roughly agree on the above might consider abandoning the Titanic before it sinks. Let's secede from the dominant no-limitations, life-destroying civilisation, and build alternative communities of human life on rational and sacred ground.

CONTRIBUTING AUTHORS

Marie Claire Brisbois

Marie Claire Brisbois is a Lecturer in Energy Policy and Co-Director of the Sussex Energy Group. Marie Claire's work explores questions of structural power, politics and influence in the contexts of energy, water and natural resource governance. Her other research interests include participatory and collaborative environmental governance, the interface between science and policy, and social mobilisation in times of transformative change.

Eileen Crist

Eileen Crist is Associated Professor Emerita in the Department of Science, Technology, and Society at Virginia Tech. She holds a Ph.D. in Sociology from Boston University, and a B.A., also in sociology, from Haverford College. Her work focuses on the ecological crisis and its root causes and pathways toward creating an ecological civilisation. She is the author of *Images of Animals: Anthropomorphism and Animal Mind* (1999) and coeditor of a number of books, including *Keeping the Wild: Against the Domestication of Earth* (2014) and *Protecting the Wild: Parks and Wilderness, the Foundation for Conservation* (2015). She is also the author of numerous academic papers and writings for general audience readers. Her most recent book, *Abundant Earth: Toward an Ecological Civilisation,* was published by University of Chicago Press in 2019. Eileen is an Associate Editor of the journal *The Ecological Citizen.*

Sibylle Frey

Sibylle Frey is an environmental scientist, consultant, and editor for the MAHB. She is a former Research Fellow at the Stockholm Environment Institute, has an MSc in nutritional science, and a PhD in environmental science from Brunel University. Sibylle is a published author who also peer-reviewed papers for the Journal of Industrial Ecology. She has 20 years of experience working on sustainable consumption and production models, as well as ten years of experience in the airline food industry. She can be reached at hello@drsibylle.com

Brittany Ganguly

Brittany Ganguly is the MAHB Communications Director and also works in California statewide mental health care programs. She has Masters degrees in Public Health and Social Work and is passionate about developing public health responses to the mental health crisis caused by human induced climate change and other existential threats and more adequately addressing women's health needs across the world.

Joe Gray

Joe Gray grew up in a quiet corner of England where much of life revolved around apples and cider. His first job, some casual summer work at the age of 16, was in an orchard. And, fittingly enough, the first green project that he became involved in centred on cider too. In the late 1990s, while in the sixth form at Newent Community School, he joined a group of students working with a forward-thinking organization known as ZERI (Zero Emissions Research and Initiatives) on a project seeking to give a second life to pomace, a waste output from the cider-making process. From there, he went on to get a BA and MA in Zoology from the University of Cambridge and, later, an MSc in Forestry from Bangor University.

Joe now spends time writing both non-fiction and fiction, the latter under the pen name Dewey Dabbar. As a passionate natural historian, he runs courses for people of all ages on various aspects of nature. And he is also a co-founder of *The Ecological Citizen* and a knowledge advisor on ecological ethics for the United Nations' Harmony with Nature programme.

Joe's first ecologically themed book, *Thirteen Paces by Four,* was published by Dixi Books in January 2021.

Peter Gray

Peter Gray, Ph.D., research professor at Boston College, is author of Free to Learn (Basic Books) and Psychology (Worth Publishers, a college textbook now in its 8th edition). He has conducted and published research in neuroendocrinology, developmental psychology, anthropology, and education. He did his undergraduate study at Columbia University and earned a Ph.D. in biological sciences at Rockefeller University. His current research and writing focus primarily on children's natural ways of learning and the life-long value of play. He a founding member of the nonprofit Alliance for Self-Directed Education and a founding board member of the nonprofit Let Grow. His own play includes not only his research and writing, but also long distance bicycling, kayaking, back-woods skiing, and vegetable gardening.

Anja Heister

Anja Heister, M.S., Ph.D. is a lifelong animal rights activist and involved in advancing rights for wild animals. Originally from Frankfurt, Germany, Anja moved to Missoula, Montana/USA in 2000, where she co-founded Footloose Montana, a nonprofit organisation working towards an end of trapping of wild animals. She worked as a campaign director for a national animal rights organisation before becoming an independent researcher and writer for social change for nonhuman animals. Anja is currently working on a book advocating Compassionate Conservation and shares her life with three adopted dogs and a human partner.

Martin Hultman

Associate Professor Martin Hultman, Chalmers University of Technology, Sweden, is widely published in energy, climate and environmental issues – especially notable are the articles 'The Making of an Environmental Hero: A History of Ecomodern Masculinity, Fuel Cells and Arnold Schwarzenegger' and 'A green fatwā? Climate change as a threat to the masculinity of industrial modernity' then the books Discourses of Global Climate Change and Ecological Masculinities. Hultman leads three research groups analysing 'gender and energy', 'ecopreneurship in circular economies', 'climate change denial' and as part of his academic work he publishes chronicles in a wide range of newspapers and gives public lectures commenting on contemporary politics.

Reed Noss

Reed Noss is a freelance writer, photographer, lecturer, and consultant in natural history, ecology, and conservation and serves as Chief Science Advisor with the Southeastern Grasslands Initiative. He was formerly Provost's distinguished research professor of biology at the University of Central Florida. He received a BS degree in education from the University of Dayton, an MS degree in ecology from the University of Tennessee, and a PhD in wildlife ecology from the University of Florida. He has served as editor-in-chief of *Conservation Biology*, science editor for *Wild Earth* magazine, and president of the Society for Conservation Biology. He is an elected fellow of the American Association for the Advancement of Science. His recent research topics include disturbance ecology; road ecology; ecosystem conservation and restoration; and vulnerability of species and ecosystems to sea-level rise. He has more than 300 publications, including eight books. His most recently published books are *Forgotten Grasslands of the South: Natural History and Conservation* (Island Press, 2013) and *Fire Ecology of Florida and the Southeastern Coastal Plain* (University Press of Florida, 2018).

Andrew Olivier

Andrew Olivier divides his time between living in the art community village of Riebeek Kasteel in South Africa, and Narara Ecovillage in Australia. He is an ambassador for GEN (Global Ecovillage Network), a GENOA representative and a founding member of GEN Australia. He was involved with setting up a transition town and a pioneer of Narara Ecovillage. Primarily a businessman and entrepreneur, he is now a board trustee for GEN.

As a board trustee, his interest is on urban-ecovillages, co-living, co-housing and new models to dealing with whole of life and how we turn cities from energy sinkholes to vibrant self sustaining communities.

He is very interested in advocacy and in reaching out to business and governments, local councils to discuss different ways of creating awareness with URGENCY.

Colin Tudge

Colin Tudge is a biologist by education and a writer by trade. Colin has written a great many articles for a great many publications and for a time was on the staff of Farmers' Weekly, then New Scientist, then BBC Radio 3. But mainly he writes books—on natural history, evolution, food and farming, and, lately, on the philosophy of science and metaphysics. He also enjoys public speaking and run courses on farming and related matters at Schumacher College in Devon.

In the early 2000s Colin coined the expression "Enlightened Agriculture", sometimes abbreviated to "Real Farming" and defined informally as "Agriculture that is expressly designed to provide everyone, everywhere, with food of the highest standard, nutritionally and gastronomically, without wrecking the rest of the world".

In 2016 Colin established the College for Real Farming and Food Culture, intended to provide the intellectual underpinning of the much-needed Agrarian Renaissance. Please see http://collegeforrealfarming.org/. His latest book, *The Great Re-Think,* was published in January 2021.

Max Winpenny

Max Winpenny is an educator and communications specialist on environmental issues. His goal is to improve how science and environmental issues are communicated. He is applying this ethos of empathetic communication to the environmental crisis through presenting, writing, and podcasting.

Max's passions include the human connection to the natural world and mental health, and he looks to improve both through effective education and communication. You can see more of Max's work on mental health at mentalk2.com.

George Wuerthner

George Weurthner, is an ecologist, longtime wildlands activist, and wilderness visionary with interests in conservation history and conservation biology.

George worked as Ecological Projects Director for Tompkins Conservation which has created more than a dozen national parks in Patagonia. Currently, he is ED of Public Lands Media which summarises scientific research, making these concepts accessible to the public. He has published 38 books dealing with natural history and environmental topics.

Karen Environmental and Social Action Network

The Karen Environmental and Social Action Network (KESAN) is a community-based, non-governmental, non-profit organization that works to improve livelihood security and to gain respect for indigenous people's knowledge and rights in Karen State of Burma, where the violence and inequities of more than 60 years of civil war have created one of the most impoverished regions in the world. KESAN's approach to the development of sustainable rural livelihoods is based on principles of democratization from below and "Free, Prior and Informed Consent." They survey, carry out capacity building and facilitate dialogue to mobilize and empower local communities, leaders, organizations and policy makers who can then make better-informed development decisions.